A TEXT BOOK OF

MANUFACTURING PROCESS - II

FOR
SEMESTER – VI

THIRD YEAR (T.Y.) B. TECH COURSE IN
MECHANICAL / MECHANICAL (SANDWICH) ENGINEERING

**Strictly According to New Revised Credit System Syllabus
of Babasaheb Ambedkar Technological University (BATU),
Lonere, (Dist. Raigad) Maharashtra,**
(w.e.f. June 2019-20)

Dr. ANAND K. BEWOOR
Ph.D. (Mech. Engg.),
Professor,
Deptt. of Mechanical Engg.,
Cummins College of Engineering for Women,
Karvenagar, PUNE.

PRALHAD A. PESODE
M. Tech. (Mfg. Tech.) NIT-Trichy,
Assistant Professor,
School of Mechanical Engg.,
MIT World Peace University (MIT WPU),
Kothrud, PUNE.

N1106

MANUFACTURING PROCESS - II (MECH. / MECH. (SAND.), BATU) ISBN : 978-93-89686-71-5

First Edition : **December 2019**
© : **Authors**

Every effort has been made to avoid errors or omissions in this publication. In spite of this, errors may have crept in. Any mistake, error or discrepancy so noted and shall be brought to our notice shall be taken care of in the next edition. It is notified that neither the publisher nor the author or seller shall be responsible for any damage or loss of action to any one, of any kind, in any manner, therefrom.

Published By :
NIRALI PRAKASHAN
Abhyudaya Pragati, 1312 Shivaji Nagar
Off J.M. Road, PUNE 411005
Tel : (020) 25512336/37/39
Email : niralipune@pragationline.com

➢ DISTRIBUTION CENTRES

PUNE

Nirali Prakashan (Local)	:	119 Budhwar Peth, Jogeshwari Mandir Lane, Pune 411002, Maharashtra
		Tel : (020) 2445 2044, Mobile : 9657703145, Email : niralilocal@pragationline.com
Nirali Prakashan (Outstation)	:	S. No. 28/27 Dhayari, Near Asian College, Dhayari, Pune 411041, Maharashtra
		Tel : (020) 2469 0204, Fax : (020) 2469 0316, Mobile : 9657703143
		Email : bookorder@pragationline.com

MUMBAI

Nirali Prakashan	:	385 S.V.P. Road, Rasadhara Co-op. Hsg. Society Ltd., Girgaum, Mumbai 400004, Maharashtra
		Tel : (022) 2385 6339 / 2386 9976, Fax : (022) 2386 9976, Mobile : 9320129587
		Email : niralimumbai@pragationline.com

➢ DISTRIBUTION BRANCHES

JALGAON

Nirali Prakashan	:	34 V. V. Golani Market, Navi Peth, Jalgaon 425001, Maharashtra
		Tel : (0257) 222 0395, Mob : 94234 91860, Email : niralijalgaon@pragationline.com

KOLHAPUR

Nirali Prakashan	:	New Mahadvar Road, Kedar Plaza 1st Floor, Opp. IDBI Bank
		Kolhapur 416012, Maharashtra. Mobile : 9850046155
		Email : niralikolhapur@pragationline.com

NAGPUR

Nirali Prakashan	:	Above Maratha Mandir, Shop No 3, Second Floor,
		Rani Jhanshi Square, Sitabuldi, Nagpur 440012, Maharashtra
		Tel : (0712) 254 7129, Email : niralinagpur@pragationline.com

DELHI

Nirali Prakashan	:	4593/15 Basement, Agarwal Lane, Ansari Road, Daryaganj
		Near Times of India Building, New DelhiV 110002 Mobile : 8505972553
		Email : niralidelhi@pragationline.com

BENGALURU

Nirali Prakashan	:	Maitri Ground Floor, Jaya Apartments, No. 99, 6th Cross, 6th Main,
		Malleswaram, Bengaluru 560003, Karnataka
		Mobile : 9449043034, Email : niralibangalore@pragationline.com

Note: Every possible effort has been made to avoid errors or omissions in this book. In spite this, errors may have crept in. Any type of error or mistake so noted, and shall be brought to our notice, shall be taken care of in the next edition. It is notified that neither the publisher, nor the author or book seller shall be responsible for any damage or loss of action to any one of any kind, in any manner, therefrom. The reader must cross check all the facts and contents with original Government notification or publications.

niralipune@pragationline.com | www.pragationline.com
Also find us on [f] www.facebook.com/niralibooks

PREFACE

It gives us great pleasure to present the book **" Manufacturing Process – II "** for the students of **Semester VI Third Year (T.Y.) B. Tech. Course Mechanical / Mechanical (Sandwich) Engineering of Dr. Babasaheb Ambedkar Technological University (BATU), Lonere, Dist. Raigad (Maharashtra).** This book is strictly as per the new revised syllabus 2019-20 Pattern, effective from the Academic Year July 2019-20.

In New Revised Syllabus, there will Class Assessment (CA) 20 Marks, Mid Sem. Exam. (MSE) 20 Marks and End Sem. Exam. (ESE) 60 Marks. End Sem. Exam. will be based on all Six units and each unit will carry 20 Marks.

The Theory Course will have 3 Credits.

The basic objective of this book is to bridge the gap between the vast contents of the reference books, written by the renowned International Authors and the concise requirements of Undergraduate Students. This book has been written in a comprehensive manner using Simple and Lucid language, keeping in mind students' requirements. The main emphasis has been given on exploring the basic concepts rather than merely the Information. Solved Examples and Exercises have been provided throughout the book and at the end of the Unit. Also, we have given **Model Question Papers** for practice at the end of book.

Our special thanks to our family members, students and all those who directly or indirectly supported us in this project.

We also take this opportunity to express our sincere thanks to Shri. Dineshbhai Furia, Shri. Jignesh Furia, Mrs. Nirali Verma, Shri. M. P. Munde and entire team of Nirali Prakashan, namely Mrs. Deepali Lachake (Co-ordinator), and her colleagues who really have taken keen interest and untiring efforts in publishing this text.

The advice and suggestions of our esteemed readers to improve the text are most welcome and will be highly appreciated.

Pune   **Authors**

SYLLABUS

Unit 1: Abrasive Machining and Finishing Operations

Introduction; Abrasives and Bonded Abrasives: Grinding Wheels, Bond Types, Wheel Grade and Structure; Grinding Process: Grinding-wheel wear, Grinding Ratio, Dressing, Truing and Shaping of Grinding Wheels, Grindability of Materials and Wheel Selection; Grinding Operations and Machines; Design Considerations for Grinding; Finishing Operations

Unit 2: Mechanics of Metal Cutting

Geometry of single point cutting tools, terms and definitions; chip formation, forces acting on the cutting tool and their measurement; specific cutting energy; plowing force and the "size effect"; mean shear strength of the work material; chip thickness: theory of Ernst and merchant, theory of Lee and Shaffer, friction in metal cutting

Unit 3: Thermal Aspects, Tool Wear, and Machinability

Temperature in Metal Cutting: Heat generation in metal cutting; temperature distribution in metal cutting, effect of cutting speed on temperatures, measurement of cutting temperatures

Tool life and tool Wear: progressive tool wear; forms of wear in metal cutting: crater wear, flank wear, tool-life criteria,

cutting tool materials: basic requirements of tool materials, major classes of tool materials: high-speed steel, cemented carbide, ceramics, CBN and diamond, tool coatings; the work material and its machinability

Cutting fluids: Action of coolants and application of cutting fluids.

Unit 4: Processing of Powder Metals

Introduction; Production of Metal Powders: Methods of Powder Production, Particle Size, Shape, and Distribution, Blending Metal Powders; Compaction of Metal Powders: Equipment,Isostatic Pressing, Sintering; Secondary and Finishing Operations; Design Considerations.

Unit 5: Processing of Ceramics and Glasses

Introduction; Shaping Ceramics: Casting, Plastic Forming, Pressing, Drying and Firing, Finishing Operations; Forming and Shaping of Glass: Flat-sheet and Plate Glass, Tubing and Rods, Discrete Glass Products, Glass Fibers; Techniques for Strengthening and Annealing Glass: Finishing Operations; Design Considerations for Ceramics and Glasses.

Unit 6: Processing of Plastics

Introduction; Extrusion: Miscellaneous Extrusion Processes, Production of Polymer Reinforcing Fibers; Injection Moulding: Reaction-injection Molding; Blow Moulding; Rotational Moulding; Thermoforming; Compression Moulding; Transfer Moulding; Casting; Foam Moulding; Cold Forming and Solid-phase Forming; Processing Elastomers.

CONTENTS

❖ ❖ ❖

UNIT I

ABRASIVE MACHINING AND FINISHING OPERATIONS

1.1 INTRODUCTION

- Grinding is the process of removal of very fine quantities of materials form the contact surface of workpiece by using grinding wheel made up of abrasive grains.

- According to the machining requirement the specific size of abrasive grain particles are mixed with the bonding material and then pressed into a disc shape of designed thickness and diameter.

- **Infinite Cutting Edges** available in grinding wheel as compared to other machining operations like drilling, milling.

- Grinding process used approximately 25% of all machining processes used for roughing and finishing processes.

- Grinding process is used as a precise finishing process to get the desired surface characteristics (Surface finish), correct size and accurate shape of the component.

- Fig. 1.1 shows the basic grinding principle, the basic arrangement of grinding which has some similarities with the up milling operation except that the cutting edges are irregularly shaped and randomly distributed.

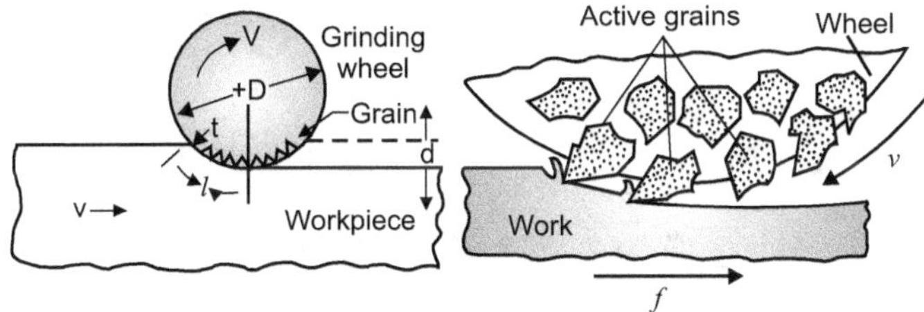

(a) Basic structure　　　　**(b) Cutting action of grains**

Fig. 1.1 : Grinding operation

- **Active Grains** take actual part in material removal; gradually the sharp edges of the active grains wear out and become blunt.

- This results in larger forces on the active grains during machining.

- When the cutting edges to blunt and the force is adequately high, the grain may either get fractured or break away from the wheel.

- The grinding wheels have self-sharpening characteristics, because when fracture take place, new sharp cutting edges are generated, and when the whole grain is removed, new grain below the layer of active grain become exposed and active.

1.2 ADVANTAGES OF GRINDING

- Large number of cutting edges on the grinding wheel creates extremely smooth finish desirable at contact and bearing surfaces can be produced only by grinding operation.

- The wheel has considerable width results no marks of feeding.

- Complex profiles can be produced accurately with relatively low-cost truing templates.

- Very little pressure is required in this process, thus permitting its use on very light work that would otherwise tend to spring away from the tool.

- Abrasives have very high hardness; are less sensitive to heat compared to other materials and can sustain high temperatures.

- Grinding is the convenient method of removing material from materials after hardening.

- Grinding unlike conventional machining need not cut through the hard skin of forgings, etc.

1.3 TYPES OF GRINDING OPERATIONS

- The common types of grinding operations are surface grinding with horizontal spindle, surface grinding with vertical spindle, external cylindrical grinding, internal cylindrical grinding, centreless grinding, form grinding.

- The basic principles of these grinding operations are shown in Fig. 1.2.

(1.1)

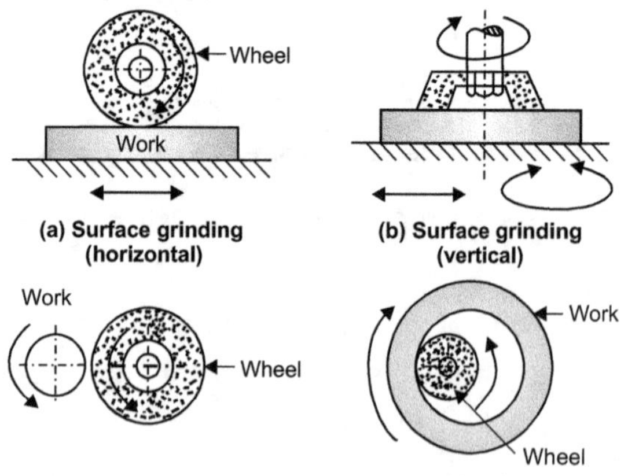

(a) Surface grinding (horizontal)

(b) Surface grinding (vertical)

(c) Cylindrical grinding

(d) Internal cylindrical grinding

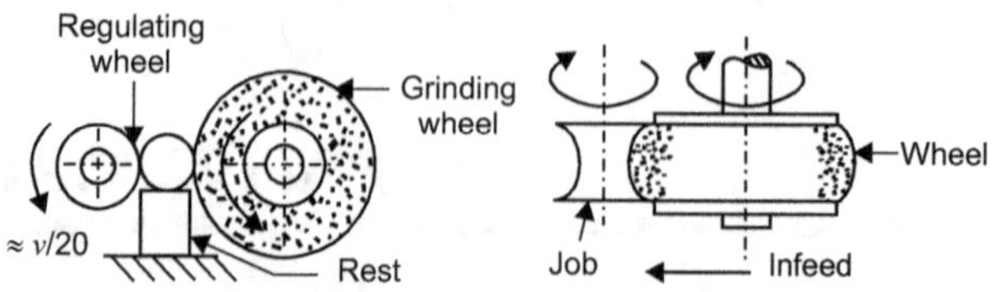

(e) Centreless grinding

(f) Form cylindrical grinding

Fig. 1.2 : Various grinding operations

1.4 TYPES OF GRINDING MACHINE

- Grinding machines are generally classified on the basis of finish parts having cylindrical, flat or internal surface.

- The Fig. 1.3 shows the broad classification of grinding machine.

Classification of Grinding Machine

Cylindrical Grinder	Internal Grinder	Surface Grinder	Tool Grinder	Special Grinder
• Centerless	• Work rotated in chuck	• Planer type or reciprocating table (Horizontal and vertical spindle)	• Universal	• Swinging frame
• Chucking	• Work rotated and held by rolls	• Rotating table (Horizontal and vertical spindle)	• Vertical Spindle	• Cut-off sawing
• Tool-post	• Work stationary			• Super finishing
• Work between centres				• Flexible shaft
				• Honing and lapping

Fig. 1.3 : Classification of grinding machine

1.4.1 Cylindrical Grinding Machine

- It is generally used for producing external cylindrical surface, which may be parallel or tapered or fillets, grooves, shoulders and other formed surfaces of revolution.

- The main sub-classification of cylindrical grinders depends upon the method of supporting the work, between the centres, or centreless.

- The grinding wheel an independent power and is driven at high speed suitable for grinding operation, both the work and grinding wheel rotate counter clockwise.

- The work rotates at much lower speed compared to that of grinding wheel.

- In case of centreless type, the work is supported by the arrangement of the work rest, a regulating wheel and the grinding wheel.

- The depth of cut is controlled by feeding wheel into the work. The depth of cut for roughing cut is normally 0.05 mm, but for finishing cuts, the feed is reduced to 0.005 mm.

- **Universal Cylindrical Grinding Machines** are very versatile; best suited for tool room applications. Both work and wheel head can be swivelled. It is possible to grind tapered surfaces and use this machine for internal, face and surface grinding by using some attachments. It is also possible to perform plunge cylindrical grinding, taper plunge grinding on it.

- The Fig. 1.4 shows the various operations carried on a universal cylindrical grinding machine.

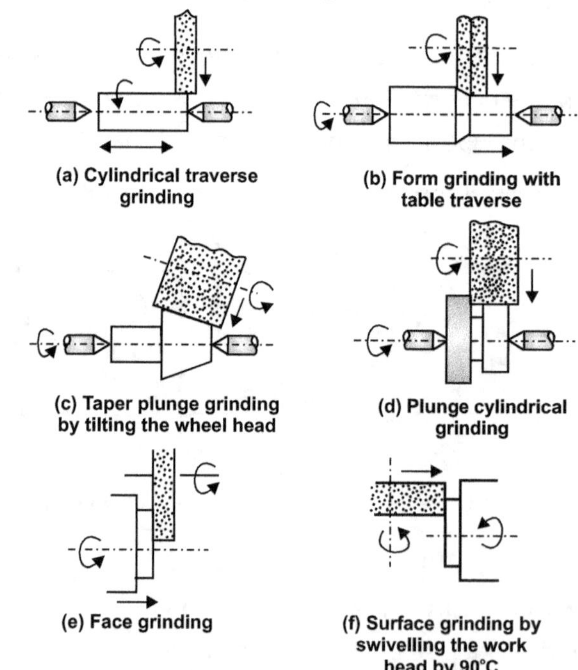

(a) Cylindrical traverse grinding

(b) Form grinding with table traverse

(c) Taper plunge grinding by tilting the wheel head

(d) Plunge cylindrical grinding

(e) Face grinding

(f) Surface grinding by swivelling the work head by 90°C

Fig. 1.4 : Various operation on a universal cylindrical grinding machine

1.4.2 Centreless Grinding Machine   [Feb. 15, 17]

- Centreless grinding has gained much popularity and this process plays an important role in the sphere of grinding.

- The work is not supported between centres but is held against the face of grinding wheel by combination of supporting rest and a regulating wheel.

- Thus centreless grinding does not require centre holes, drivers and other fixtures for holding the workpiece.

- During the process, the workpiece is supported on a work rest blade and the regulating wheel holds the workpiece against the horizontal force of action controlling its size and imparting the necessary rotational and longitudinal feed.

- Fig. 1.5 shows the principle working of centreless grinding. In the side view it is shown that the regulating wheel is inclined at angle θ to the grinding wheel.

- The rotation of wheel is such that the workpiece rotates in downward direction as this position can be balanced by the work-rest.

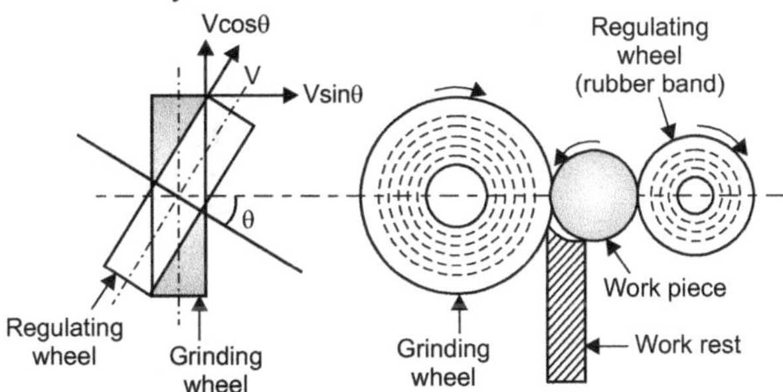

Fig. 1.5 : Basic structure of centreless grinding

There are three methods of centreless grinding are as follows :

1. Through-feed
2. In-feed
3. End-feed

1. Through-Feed

- This is simplest method and is applied only to plain parallel parts (Roller pins and straight long bars) which are difficult to grind by ordinary cylindrical grinding method.

- In this case, the gap between the regulating wheel and grinding wheel is adjusted equal to the desired diameter of the workpiece and then the job is fed and passed through the wheels.

- The machine can remove upto 0.38 mm of material on the diameter in one pass. For finish grinding, high speed of regulating wheel combined with less inclination are recommended.

- The axial movement of the work past the grinding wheel is obtained by tilting the regulating wheel downward at the feeding end, at a slight angle from horizontal.

- Due to this inclination, the peripheral speed of the regulating wheel is resolved into two components work rotation speed, $V_w = V_r.\cos\theta$ and rate of feed $F = V_r \cdot \mu \sin\theta$.

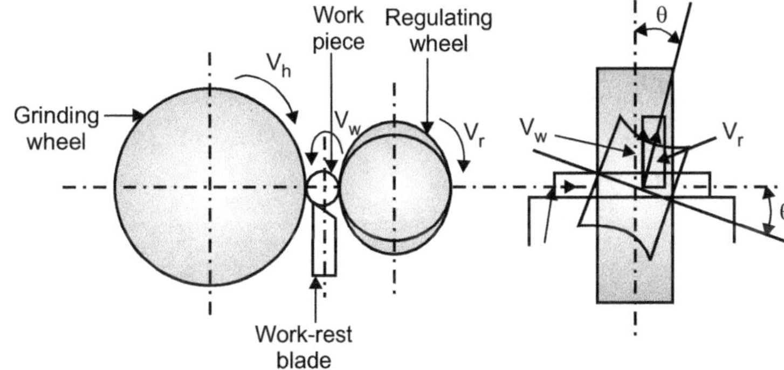

Fig. 1.6 : Through-feed method

- Where μ is the coefficient which accounts for slipping of the work with respect to the grinding wheel (μ = 0.94 to 0.98).

- Angle θ is usually taken from 1° to 5°; feed rate mainly depend on the angle θ.

Advantages

- The component can be loaded and unloaded from the machine rapidly, so grinding is almost continuous for through feed grinding.

- The regulating wheel and work-rest blade practically eliminates any deflection of the workpiece. This permits maximum material removal rates.

- Minimum wear in the large grinding wheels, this minimises the adjustments needed for staying within dimensional tolerances and maximises the periods of time between wheel dressing.

- Less grinding allowance needed, because the out-of-roundness is corrected across the diameter rather than the radius.

Disadvantages

- Setup time is maximum for a centreless grinding operation.

- It may be necessary to have special equipment and additional setup time for special profiles, so it is useful only for large volume production.

- It is not suitable for large size of workpiece.

1.4.3   Internal Centreless Grinding Machine

- This process is used to grinding of hollow ring type of component.

- The basic arrangement of internal centre less grinding shown in Fig. 1.7; the workpiece loaded in between support, pressure and regulating rollers mounted in housing.

- The grinding wheel and the workpiece rotate in the same direction while the regulating wheel in the opposite direction.

- The regulating wheel is tilted at a small inclination angle to control the feed of the work past the grinding wheel. Because of the need to support the grinding wheel, through feed of the work as in external centerless grinding is not possible.

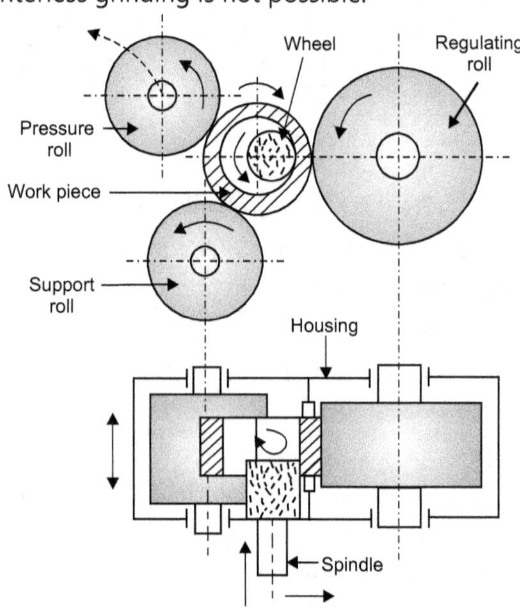

Fig. 1.7 : Internal centreless grinding

- Therefore this grinding operation cannot achieve the same high-production rates as in the external centerless process.

- It's advantage is that it is capable of providing very close concentricity between internal and external diameters on a tubular part such as a roller bearing face.

1.4.4   Surface Grinding Machine

- Plain or flat surface ground on surface grinding machine.

- By using special fixtures and form dressing devices useful for grinding of angular or formed surface.

- For holding of the workpiece various type of attachment required such as magnetic chuck for smaller component, and for large size of component clamping done with the help of pads, strap.

- **Horizontal Spindle Type**, the face of the wheel is used for grinding. The work is traverse under the wheel face gradually. Work is also feed laterally at each end of the stroke so that a required area may be ground.

- **Rotary Table Types**, with horizontal spindles are generally used for accurate grinding. The workpiece is mounted on a spinning horizontal axis table and the grinding wheel spindle is conceded on a wheel slide which can be travel across the workpiece.

- **Vertical Spindle or Cup Grinding**, flat surfaces are generated with the side of wheel mounted on the vertical spindle. The wheel remains fixed and is feed down or up, until the work is finished.

- Horizontal spindle and rotating table

- Vertical spindle and rotating table

- Horizontal spindle and reciprocating table

- Vertical spindle and reciprocating table

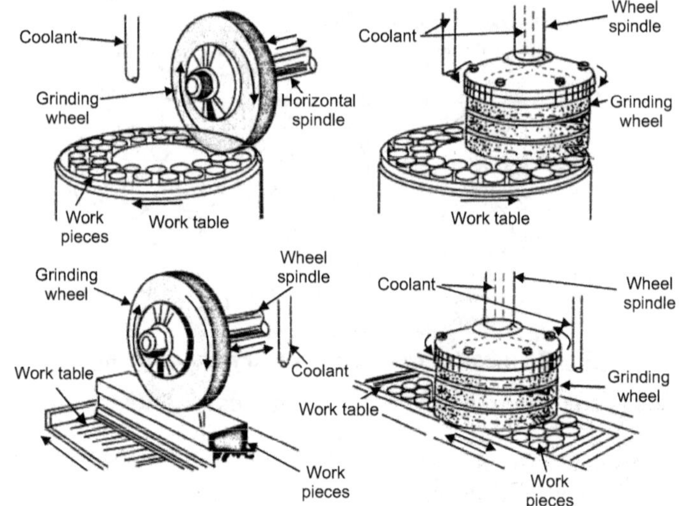

Fig. 1.8 : Surface grinding methods

1.4.5   Tool and Cutter Grinding Machine

[Nov. 16, 17; May 17]

- In offhand grinding, a bench or pedestal-type grinder is used.

- The tool is held in hand and is moved across the face of the wheel continuously to avoid extreme grinding on one spot.

- In mass production, where a number of tools of identical shape and size, or the tools requiring accuracy and where various rake and other angles are to be set on the tool properly for economic reasons, special single purpose grinders are used.

- Special drill or tool-bit grinders are acceptable by the large amount of grinding work necessary to keep production tools in proper cutting conditions.

- In addition to that, the tools can be ground uniformly and with accurate cutting angles.

- For the sharpening of various cutters, universal type grinder is used. It is equipped with universal head, vice, head stock, tail stock and many other attachments for holding tools and cutters.

- This grinder is designed for cutter sharpening, cylindrical, taper, internal and surface grinding operations.

- The machine consists of heavy box type base, saddle directly mounted on it, table have freedom to move longitudinally on lower table.

- The head stock and tailstock are mounted on either side of the table and hold the work in between centre. The column supporting the wheel head is mounted on the base on the back of the machine.

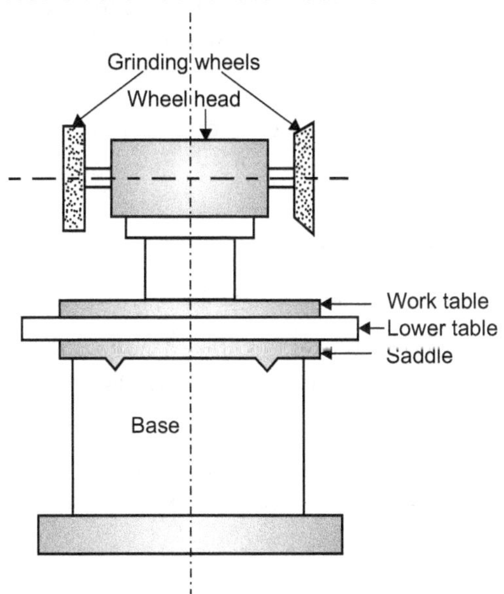

Fig. 1.9 : Tool and cutter grinder

1.4.6 Internal Grinding Machine

- This machine used for finishing internal bores and tubes, which are generally tapered and those having more than one diameter.

- It is used on production parts that have not been heat treated to save the reaming cost.

- The wheel is rotated in a set position while the work is slowly rotated and traversed, back and forth.

- The wheel is rotated and at the same time reciprocated back and forth through the length of hole. The work is rotated slowly. This type of grinder is also called chucking grinder.

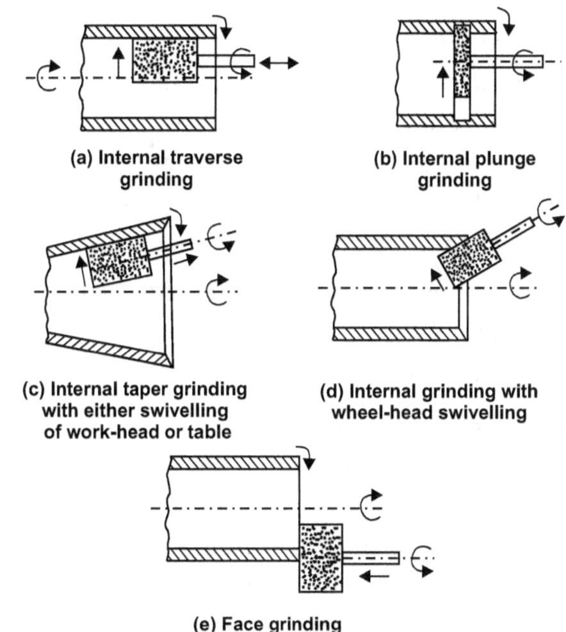

Fig. 1.10 : Various internal grinding operations

GRINDING WHEEL

1.5 PERFORMANCE PARAMETERS OF A GRINDING WHEEL

1.5.1 Abrasive Type

- Grinding wheels are made of abrasive grain particles bonded together by means of some suitable bond.

- An abrasive is a hard material which can be used to cut or wear away other materials. When fractured, it forms sharp cutting edges and corners.

Abrasive Particles Used for Grinding Wheels are :

1. Natural abrasives
2. Artificial abrasives

1. Natural Abrasives

- These are produced by uncontrolled forces of nature. The following are the generally found and used natural abrasives:

  ➢ Sand stone or solid quartz
  ➢ Emery (50-60% crystalline Al_2O_3 + Iron oxide)
  ➢ Corundum (75.90% crystalline Al_2O_3 + Iron oxide)
  ➢ Natural Diamonds, the abrasive grains called "borts" are produce by crushing unsuitable gem stones to the required particle size. Their use is limited because of the tendency of dulling and glazing of the cutting faces.
  ➢ Garnet

- Natural abrasives lack standardization of proper and reliability and have been largely replaced by artificial or manufactured abrasives.

2.   Artificial Abrasives

- The quality and composition of artificial particles can be easily controlled and their efficiency is better than that of natural abrasives, because of abrasives grains particles have well defined and controlled properties of hardness, toughness and type of structure.

Most Commonly Used Artificial or Manufactured Abrasives are :

1.   Aluminium Oxide (Al$_2$O$_3$)

- The trade names for Al$_2$O$_3$ are "Alundum"; "Aloxite" and "Borolon", composed primarily of crystalline aluminium oxide.
- It is obtained by melting bauxite ore which is mainly aluminium hydroxide with this oxide in an electric furnace, impurities remove by adding iron chip and coke, and then refined aluminium oxide comes out.
- It is crushed and rolled into small grains, treated magnetically to remove ferrous impurities and washed.
- Pure Al$_2$O$_3$ grit with voids (structure defect) leads to unusually sharp free cutting action with low strength, which is better in fine tool grinding operation, and heat sensitive operations on hard, ferrous materials.
- Brown aluminium oxide (doped with TiO$_2$) possesses lower hardness and higher toughness than the white Al$_2$O$_3$ and is recommended heavy duty grinding to semi finishing.

2.   Silicon Carbide (SiC)

- Silicon carbide is a chemical compound of silicon and carbon; it is produced by fusing quartz sand and powdered coke in an electric furnace. At temperature around 2300°C, the silicon of sand combines with carbon of coke to form silicon carbide, sawdust and slat help to remove impurities.
- The core of loosely join silicon carbide crystals is broken into individual grains.
- Two types of silicon carbide : Pure silicon carbide of green variety and black or grey.
- Black carbide containing at least 95% SiC is less hard but tougher than green SiC and is efficient for grinding soft nonferrous materials. Green silicon carbide contains at least 97% SiC. It is harder than black variety and is used for grinding cemented carbide.
- Silicon carbide is harder than alumina but less tough; it is also poorer to Al$_2$O$_3$ because of its chemical reactivity with iron and steel.

3.   Diamond

- Natural diamond grains is characterized by its random shape, very sharp cutting edge and free cutting action and is exclusively used in metallic, electroplated and brazed bond.
- Diamond is the hardest material that can be used of cutting hard tool material, Knoop hardness of diamond is approximately 8000 kg/mm^2.
- Synthetic diamonds; a form of pure carbon is mainly used for truing and dressing other grinding wheels, for sharpening carbide tools, and for processing glass, ceramics and stone.
- Diamond grit is best suited for grinding cemented carbides, glass, sapphire, stone, granite, marble, concrete, oxide, non-oxide ceramic, fibre reinforced plastics, ferrite, graphite.

4.   Boron Carbide

- Boron carbide is a mixture of boron and carbon (B$_4$C).
- The grains are characterized by sharp faces and a higher abrasive ability, and the grain size not over 120 μm.
- Boron Carbide is mainly used for grinding and lapping very hard metals, hard alloys, glass and gems (ruby, topaz).

5.   Cubic Boron Nitride (cBN)

- Cubic boron nitride is a mixture of boron and nitrogen.
- It is used for grinding workpiece of tool and die steels as well as many types of hard high-alloy steels.
- Cubic boron nitride (cBN) is the second hardest material; because of its chemical stability is the abrasive material of choice for efficient grinding of HSS, alloy steels, HSTR alloys.

Grain Size

- For finishing and polishing operations, abrasive materials are crushed to obtain particles of desired size as "grains" or "grits".
- The crushing is continued until the abrasive becomes a fine powder known as Flour required for various finishing operation.
- The grain size of the abrasive material of which the grinding wheel is made influences the metal removal rate and the quality (surface finish) of a workpiece.
- Grain size is denoted by a number; indicating the number of meshes per square inch of the sieve through which the grains pass when they are graded after crushing.

- A grain size of 70 means that the abrasive will pass through a sieve having 70 openings per square inch.

Standard Grain Sizes	
Coarse	10, 12, 14, 16, 20, 24
Medium	30, 36, 46, 54, 60
Fine	70, 80, 90, 100, 120, 150, 180
Very fine	220, 240, 280, 320, 400, 500, 600, 700, 800, 900, 1000

- The fine grains have a very small depth of cut and less heat generation machining features that gives a faster metal removal and good surface finish is obtained.

- Coarse grains are better for higher stock removal rate in roughing operation.

1.5.2 Bonding Materials

- A bond is a material that holds the abrasive grains together firmly the mixture to be kept in a desired shape in the form of the grinding wheel.

- Bonds must be adequately strong to with stand the stresses of the high speed rotating grinding wheel.

- They must be capable of holding the abrasive grains strongly, yet must not be so dense as to slow down the cutting action.

Most Common Type of Bond Materials are as follows :

1. Vitrified Bond

- The most extensively used bond is the vitrified bond; suitable for high stock removal even at dry condition.

- It is the mixture of clay, fluxes (feldspar) and quartz, which hardens to a glass-like structure on heating to a temperature at 1250°C, which develop its strength.

- This bond is strong, highly porous, resistant to heat, highly chemically stable (not affected by water, oils and acids), has high cutting capacity and properly removes heat.

- Vitrified bonded wheels can be used with a circumferential speed upto 35 m/s.

- The disadvantage of vitrified-bonded wheels is their brittleness; it cannot be used where mechanical impact or thermal variations are like to occur, also not recommended for very high speed grinding because of possible breakage of the bond under centrifugal force.

2. Silicate Bond

- Sodium silicate is mixed with water glass (NaSiO$_3$) mixed with zinc oxide, lime; mixture is moulded in a mould, dried for definite time and finally heated to a temperature of about 270°C for about 20-80 hours.

- It is not as strong as vitrified bond, because bring out of dull grains occurs more intensively.

- The wheels become soft when to moisture, affected by dampness, but less sensitive to shock.

- The friction in grinding action is less, due to this less heat generation; these will are more suitable for grinding cutters, blades, knives, precision tools etc.

3. Resinoid or Synthetic Resin

- A resinoid bond is a thermosetting resin or plastic such as phenol formaldehyde.

- A resinoid bond are very strong and elastic, highly stable under variable loads as compare to vitrified bond.

- The resinoid bond porosity lower than that of vitrified-bonded wheels, however it is not heat and chemical resistant.

- Conventional abrasive resin bond have vibration absorbing characteristics therefore they are widely used for heavy duty grinding because of their ability to withstand shock load.

- Resin bond is not suggested with alkaline grinding fluid for a possible chemical reaction leading to bond weakening.

- Fiberglass reinforced resin bond is used with cut off wheels which requires added strength under high speed operation.

- These wheels can be readily reinforced with steel rings, or fibreglass or other fibers to increase their flexural strength.

4. Rubber Bond

- This bond consist of pure rubber with vulcanizing agent(sulphur) and fillers.

- The abrasive grains are spread between rubber sheets and they are then rolled to the desired thickness and finally vulcanised.

- Due to vulcanisation, the whole mixture becomes joined and acts as a solid wheel, rubber act as a bond.

- The resulting grinding wheels are stronger and may be rotated at higher rotational speeds.

- Wheels as thick as 0.1 mm can be, therefore, most suitable for fine parting off operation, operational speed 3000 – 5000 m/min, rubber bond was once popular for finish grinding on bearings and cutting tools.
- These wheels are also used as cutting off a regulating wheels in centreless grinding.

5. Shellac Bond

- The abrasive particles are coated with shellac and mixture heated to achive uniform mixing and then pressed and rolled into the desired shapes.
- This bond is chiefly used for making strong, thin wheels having some elasticity, at one time this bond was used for flexible cut off wheels, but present use of shellac bond is limited to grinding wheels engaged in fine finish of rolls.
- These wheels gives high surface finish and are used for grinding parts like camshafts and mill rolls.

6. Metallic Bond

- This is used in the making of diamond and cubic boron nitride wheels.
- Copper, tin, aluminium and their alloys are used as bonding materials, as a high thermal conductive materials of wheels.
- Metallic bond is widely used with super abrasive wheels.
- Extremely high toughness of metallic bond wheels makes these very effective in those applications where accuracy as well as higher material removal is desired.

7. Brazed Bond

- Brazed bond is a recent development, allows crystal contact as high 60-80%.
- In addition grit spacing can be precisely controlled.
- This bond is particularly suitable for very high stock removal either with diamond or cBN wheel.

1.5.3   Grade

- It is also called the hardness of the wheel and designates the force holding the grains.

- The grade of a wheel depends on the kind of bond, structure of wheel and amount of abrasive grains.
- Greater bond content and strong bond results in harder grinding wheel.
- Harder wheels hold the abrasive grains till the grinding of wheel increases to a great extent. The grade is denoted by letter grades as;

Soft	Medium	Hard
ABCDEFGH	IJKLMNOP	ORSTUVWXYZ

- Soft wheels are normally used for hard materials, and hard wheels are used for soft materials.
- While grinding hard materials, the grit is likely to become dull quickly, thereby increasing the grinding force, which tend to knock off the dull abrasive grains.

1.5.4   Structure

- The structure of a grinding wheel represents the grain spacing.
- It has two main type open or dense, generally denoted by numbers.
- The spacing between the grains allow for chips to collect.
- This helps avoiding the loading of the grind wheel.
- Open structures are used for high material removal and consequently produce rough finish.
- Dense structures are used for precision forms and profile grinding.

1.5.5   Designation

- All grinding wheels have some marking, in the form of conventional designation are applied with an indelible paint to the wheel flat surface.
- The grinding wheel making system should be able to specify the abrasive used, grain size, grade, structure and bond used in the sequence.
- A typical grinding wheel designation shown in Fig. 1.11.

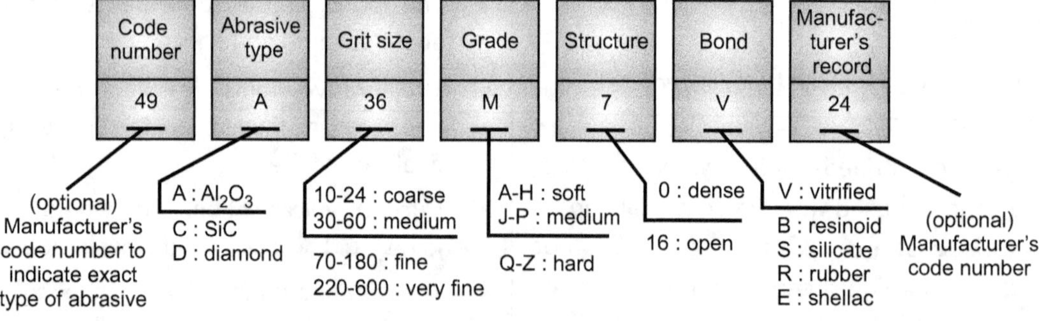

Fig. 1.11 : Designation of grinding wheel

1.5.6 Grinding Wheel Wear [May 16]

- **Grain Fracture:** A portion of the grain breaks off, but the rest of the grain remains bonded in the wheel.
- **Attritious Wear:** Dulling of the individual grains, resulting in flat spots and rounded edges.
- **Bond Fracture:** The individual grains are pulled out of the bonding material.

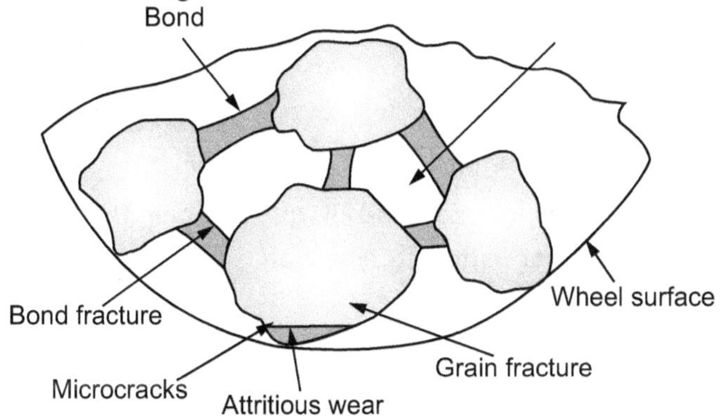

Fig. 1.12

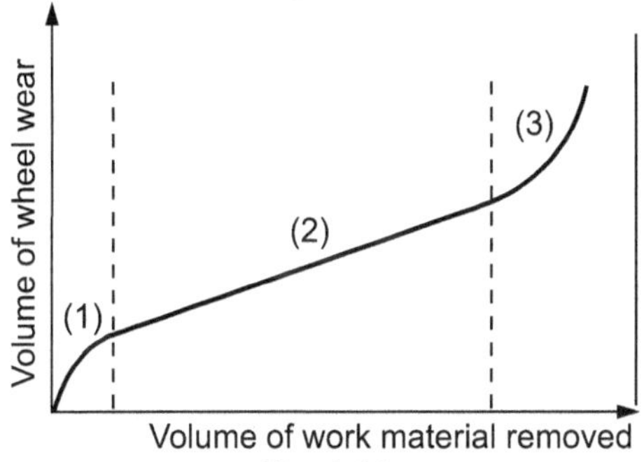

Fig. 1.13

1 : The grains are initially sharp, and wear is accelerated due to grain fracture.

2 : Characterized by attritious wear, with some grain and bond fracture.

3 : The grains become dull and the amount of ploughing and rubbing increases relative to cutting.

- **Grinding Ratio**

$$G = \frac{\text{Volume of material removed}}{\text{Volume of wheel wear}}$$

Vary greatly (2-200 or higher) depending on the type of wheel, grinding fluid, and process parameters Higher forces decrease the grinding ratio

1.5.7 Grinding Wheel Selections

Considering to select a suitable specification of grinding wheel

1. The Material to be Ground and its Hardness

- **Abrasive :** Aluminum oxide for steel and steel alloys. Silicon carbide for cast iron, non-ferrous and non-metallics.

- **Grit Size :** Fine grit for brittle materials. Coarse grit for ductile materials.

- **Grade :** Hard grade for soft materials. Soft grade for hard materials.

2. The amount of stock to be removed and the finish required

- **Grit Size :** Coarse grit for rapid stock removal as in rough grinding. Fine grit for high finishing.

- **Bond :** Vitrified for precision cutting. Resinoid and Rubber for high speed cutting.

3. Wet or dry

- **Grade :** Wet grinding, as a rule, permits use of wheels at least one grade harder than that of dry grinding without danger of burning the work.

4. The Wheel Speed

- **Bond :** Standard vitrified wheels are not exceeding 2,000mpm, for higher speeds are up to 3,600mpm.

  Standard organic bonded wheels(Resinoid, Rubber or Epoxy) are used of most applications over 2,000mpm up to 6,000mpm.

Note : Do not exceed the safe operating speed shown on a wheel tag or blotter.

5. The contact area of grinding

- **Grit Size :** Coarse grit for large contact area. Fine grit for small contact area.

- **Grade :** The smaller contact area, the harder wheel.

1.5.8 Common Grinding Wheel Shapes [May 15]

- The most common shapes of grinding wheels are shown in Fig. 1.14 (a). Grinding wheels are made in many shapes and sizes to adapt them for use in different of grinding machines for performing different classes of work.

- Shapes of the wheels have been standardized; the more common are straight-side wheels, cylinder wheels, cup wheels and dish wheels.

- Size of wheels can also be referred to system of key letters so that their dimensional specifications may be written.

- Among the grinding wheels, straight wheels are generally used for cylindrical, internal, centreless and surface grinding.

- Wheel diameter and width vary greatly depending on the class of work and machine power.

- Tapered wheel is used for grinding flat surfaces using its face or end, whereas the flaring cup wheel is used in tool room. Saucer wheel is used for sharpening circular or band saws.

- Segmented wheels are used on vertical spindle, rotary and reciprocating table surface grinders and way grinders.

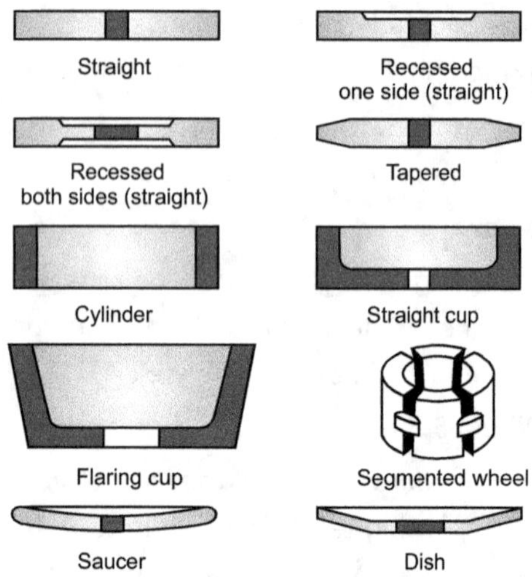

Fig. 1.14 (a) : Standard grinding wheel shapes

- For grinding special contours, grinding wheels of straight wheel type are available with large variety of faces such as flat, pointed, convex, concave, etc.

- Mounted wheels and points (Fig. 1.14 (b)) are small grinding wheels (diameter less than 50 mm) mounted firmly on to a steel spindle, mandrel or shank.

- They are made in various different shapes and sizes so as to enable grinding even in those places which are not easily accessible otherwise. They are used mostly on portable grinders.

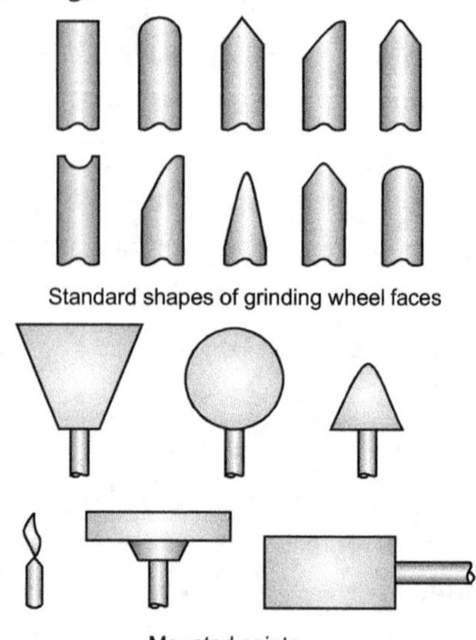

Fig. 1.14 (b) : Mounted wheels and points

1.5.9 Selection   [May 16]

The selection of the grinding wheel is based on the following elements :

- Size and shape of grinding wheel

- Type of abrasive

  These are denoted by :

  A – for Al_2O_3

  C – for SiC

  WA – for white Al_2O_3

  GC – for green grit SiC

- Grains size, it is denoted by grit number, The various numbers for different types of grain size are given below :

  ➢ Coarse grain : 8, 10, 12, 14, 16, 24

  ➢ Medium grain : 30, 36, 46, 54, 60

  ➢ Fine grain : 80, 100, 120, 150, 180

  ➢ Very fine grain : 220, 240, 280, 320, 400, 500, 600

- Grade of bond, the following classification use for grade :

  ➢ A – E : Very soft,

  ➢ G – K : Soft,

  ➢ L – O : Medium,

  ➢ P – S : Hard,

  ➢ T – Z : Very hard

- Structure, it is denoted by number from 1 to 15.

  ➢ 1 – 8 : Dense structure

  ➢ 9 – 15: Open structure

- Bond material, the following notations are followed :

  ➢ V – Vitrified

  ➢ B – Resinoid

  ➢ BF – Resinoid reinforced

  ➢ R – Rubber

  ➢ RF – Rubber reinforced

  ➢ E – Shellac

  ➢ S – Silicate

  ➢ Mg – Magnesia

- Application of grinding wheel

- Miscellaneous (wheel speed, work speed, materials to be grind)

- Wheel Speeds, grinding-wheel speeds have been fairly well established, and most machines can be adjusted to attain the desired cutting efficiency. Major problem in the past has been with the internal grinding where proper peripheral wheel speeds were difficult to obtain. Today, however, high-cycle, direction-driven wheel heads are rapidly supplanting previous types. Efficient peripheral wheel speeds, observing manufacturers' maximum limits for specific wheels, are generally as follows :

Type of Grinding	Speed, m/s
Surface grinding	20.0 – 25.0
Internal grinding	10.0 – 30.0
Diamond wheels	25.0 – 32.5

- Work Speeds, for obtaining the desired finish as well as maximum stock removal commensurate with good wheel life, proper work speeds are necessary. At the line of cutting, the work and the abrasive wheel pass in opposite directions except in internal centreless grinding. Work speed at this point should be set according to the material being ground. The low end of this range is used for roughing cuts and the high end for finishing.

Material	Speed, m/s
Soft Steel	0.15 – 0.25
Hard Steel	0.35 – 0.50
Cast Iron	1.00 – 2.00
Aluminium	0.50

- Stock Removal, most efficient stock removal is obtained with coarse-grit wheels, and while surface finish is distinctly rougher, commercially acceptable dimensional accuracy can be maintained. The following stock removal per single pass indicates the comparative rates, though these will vary somewhat for different materials : Rough grinding 0.05 – 0.50 mm, Semi-finish grinding 0.025 -.025 mm and Finish grinding 0.0125 – 0.125 mm.

1.6 MOUNTING OF GRINDING WHEELS

- Nature of mounting of a grinding wheel is shown in Fig. 1.15.
- It is clamped by flange directly on the spindle, thick compressible washers made of cardboard, rubber or leather etc. should be inserted in between the flanges and the wheel.

- The speed of grinding wheel is high; therefore any out of true part produce unbalance forces which lead to vibration. So that proper mounting of the grinding wheel is essential for the good machining conditions.

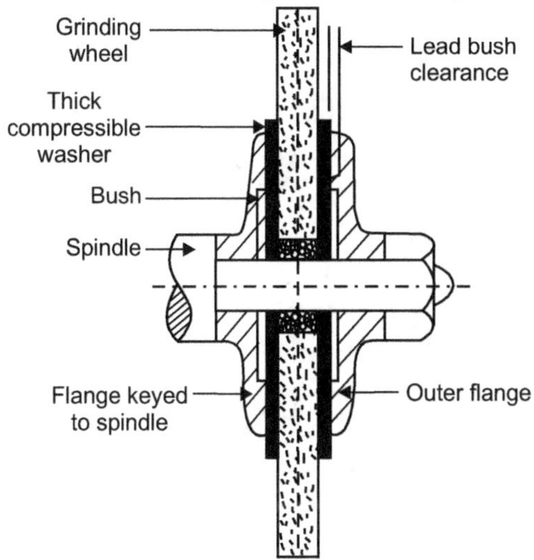

Fig. 1.15 : Mounting of grinding wheel

1.7 BALANCING AND DRESSING OF GRINDING WHEELS   [Feb. 15, 16]

1.7.1   Wheel Balancing

- When we use new grinding wheel, it should be properly balanced.
- Balance of a grinding wheel also depends upon the machine spindle as well as the condition of tightening which will have to be properly taken care of.
- For the high rotational speeds, any unbalance left would be harmful for the machine part and also produces poor surface finish.
- Such wheels are provided with movable balance weights for adjusting the balance mass location.
- The balancing operation can be done in two ways : (1) Static balancing and (2) Dynamic balancing.
- Static balancing, the grinding wheel is rotated on an arbour and the balance weights adjusted until the wheel no longer stops its rotation in any one specific position.
- The wheel is allowed to rotate such that the heavier segment of the wheel settles at rest. Place a chalk mark at the bottom-most point.
- Try to rotate the wheel slightly and see where the wheel is resting. The chalk mark should always point to the bottom, which confirms that the heaviest portion is identified.

- Two weights are now inserted such that they are equidistant from the heavy mark and slightly above the horizontal mark in that position, as shown in Fig. 1.16 (a).

- If the wheel stops again at the same point, then move the weights closer.

- It should be possible to find a point with proper balance by repeating this process, if it is not possible by any combination to find a balance, then add more balance weights as shown in Fig. 1.16 (b).

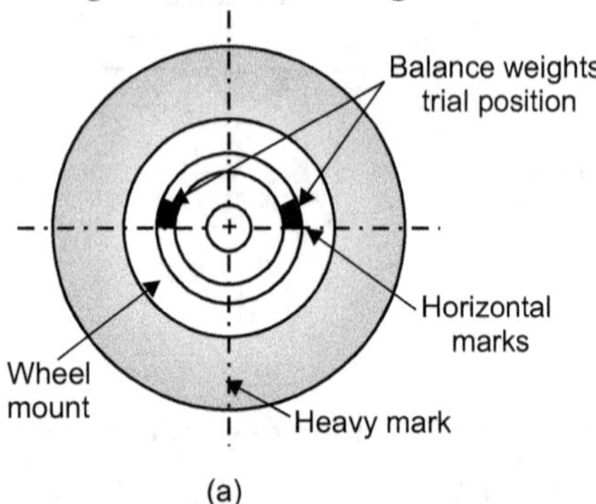

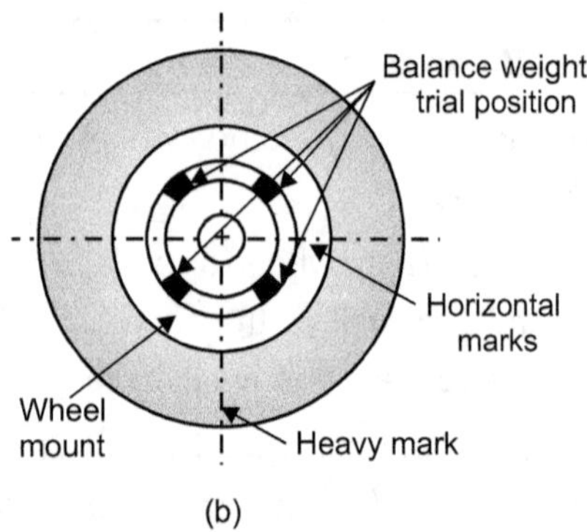

Fig. 1.16 : Balancing of grinding wheel

1.7.2   Truing and Shaping of Grinding Wheel

- Major and inevitable defects in grinding are glazing of grinding wheels. After the continuous use grinding wheel becomes dull or glazed. Glazing of the wheel is a condition in which the face or cutting edge acquires a glass like appearance.

- That is, the cutting points of the abrasives have become dull and worn down to bond. Glazing makes the grinding face of the wheel smoother and that stops the process of grinding. Sometimes grinding wheel is left 'loaded'. In this situation its cutting face is found being adhering with chips of metal.

- The opening and pores of the wheel face are found filled with workpiece material particals, preventing the grinding action. Loading takes place while grinding workpiece of softer material.

Dressing: If the grinding wheels are loaded or gone out of shape, they can be corrected by dressing or truing of the wheels.

- Dressing is the process of breaking away the glazed surface so that sharp particles are again presented to the work.

- The common types of wheel dressers known as "Star" - dressers or diamond tool dressers are used for this purpose.

- A star dresser consists of a number of hardened steel wheels on its periphery.

- The dresser is held against the face of the revolving wheel and moved across the face to dress the wheel surface. This type of dresser is used particularly for coarse and rough grinding wheels.

- For precision and high finish grinding, small industrial diamonds known as 'bort' are used.

- The diamonds are mounted in a holder. The diamond should be kept pointed down at an angle of 15° and a good amount of coolant is applied while dressing. Very light cuts only may be taken with diamond tools.

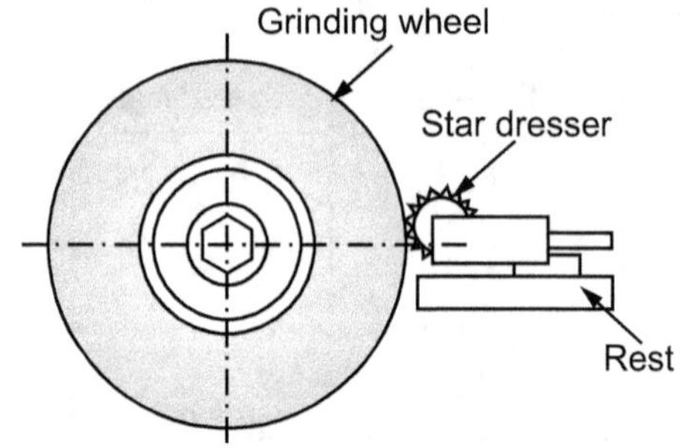

Fig. 1.17 : Dressing of a grinding wheel (Star wheel method)

Truing: The grinding wheel becomes worn from its original shape because of breaking away of the abrasive and bond.

- Sometimes the shape of the wheel is required to be changed for form grinding.

- For these purposes the shape of the wheel is corrected by means of diamond tool dressers.

- This is done to make the wheel true and concentric with the bore or to change the face contour of the wheel.

- This is known as truing of grinding wheels.

- Diamond tool dressers are set on the wheels at 15° and moved across with a feed rate of less than 0.02 mm.

- A good amount of coolant is applied during truing.

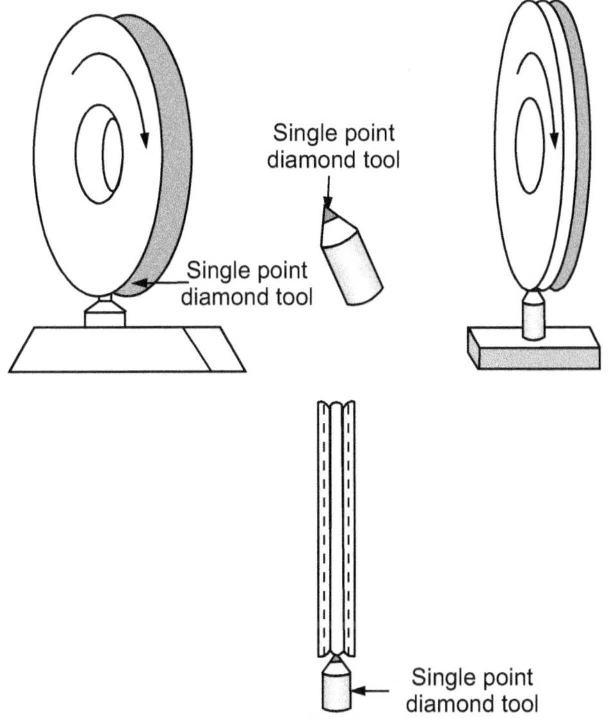

Fig. 1.18 : Application of single point diamond truing tool

Grindability of Materials :

- Grindability as a material characteristic must indicate the suitability for size reduction. Its value has to give the result of the grinding operation as a result of unit energy expenditure. And, here we meet a difficulty: the result of the operation is a new particle size distribution which cannot be described by a single value. Therefore the grindability cannot be charaterized as a difference or quotient of two numbers.

- In industrial practice, however, as a convention according to Lehmann and Haese grindability is defined as the relation of specific surface increase to energy expense. The customary units of measure for grindability are cm^2/J or cm^2/kWh.

- Generally speaking, grindability is not an absolute but a fineness dependent characteristic. With growing fineness, grindability decreases. Data in manuals are related to moderate grinding fineness, e.g. 10% residue on 0.09 mm sieve or 3000 cm^2/g Blaine surface. Industrial practice therefore simplifies further the problem.

- For a known grinding process (e.g. cement grinding in a ball mill to a grinding fineness of 3000 cm2/g Blaine). It is almost superfluous to mention that the grinding equipment influenc'es these values too. So, for example, for cement raw meal grinding. different. values will be obtained from grinding in a ball mill. in an autogeneous or in a roller mill.

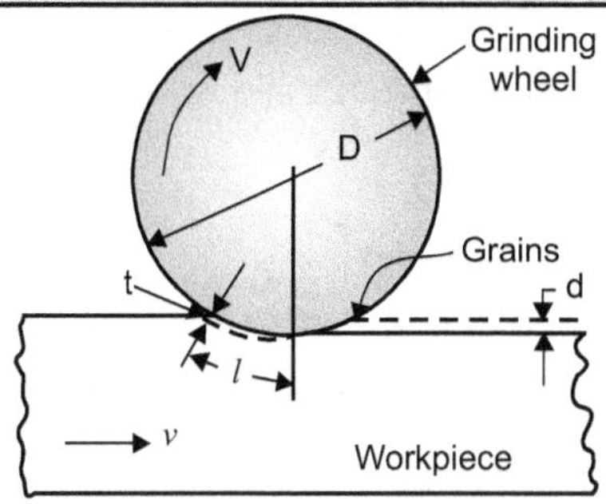

Fig. 1.19

Undeformed chip thickness t,

$$t = \sqrt{\frac{4v}{Vcr}}\sqrt{\frac{d}{D}}$$

where,   C = Number of cutting points per unit area of periphery of the wheel and is estimated in the range of 0.1 to 10 per mm^2

      r = Ratio of chip width to average undeformed chip thickness. It has approximate value of between 10 and 20.

1.9 MACHINING TIME (CALCULATION FOR CYLINDRICAL AND PLUNGE GRINDING

Grinding Time : The surface grinding operation with a horizontal axis grinding machine is shown in Fig. 1.20.

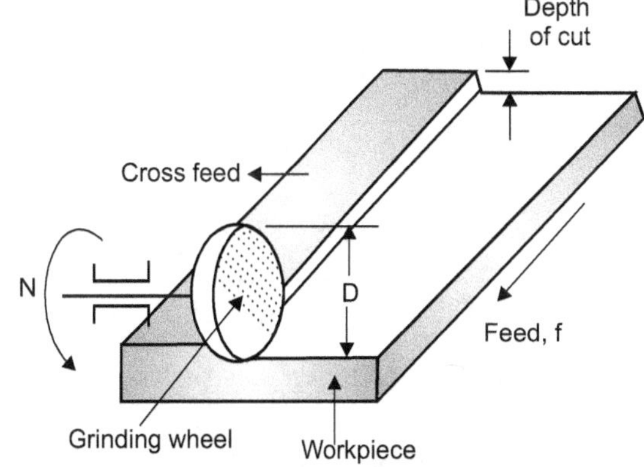

Fig. 1.20 : Grinding operation

- The grinding wheel will have to go over actual workpiece by a distance termed as the approach allowance.

- A_a which is given by

$$A_a = \sqrt{\left(\frac{D}{2}\right)^2 - \left(\frac{D}{2} - b\right)^2} = \sqrt{b(D-b)}$$

- The above value is very small since the depth of cut, b, is very small in grinding.

- However, to allow for the table reversal at each end of the table stroke, the radius of the grinding wheel is assumed as the approach allowance. Thus,

$\therefore$ Time for one pass (t) $= \dfrac{l + D}{f_t}$

where, l = Length, D = Diameter, f_t = Table feed rate

$\therefore$ Number of passes required $= \dfrac{w}{f_i}$

where, w = width, f_i = Infeed rate

Machining Time :

- **For Plunge Grinding**

Machining time $= \dfrac{\text{Machining allowance}}{\text{radial feed rate}}$ (mm/rev)

Radial feed rate = Longitudinal feed rate (mm/rev) $\times$ Speed of workpice (rpm)

- **For Cylindrical Grinding**

Machining time $= T_m = \dfrac{L \times n \times k}{N \times f}$

where,

L = length to be ground,

n = no. of passes

k = coefficient for accuracy and surface finish (1 to 1.6)

$\quad$ = 1 to 1.2 for rough grinding

$\quad$ = 1.3 to 1.7 for finish grinding

f = longitudinal feed rate (mm/rev.)

N = speed of workpiece

Number of passes (n) $= \dfrac{\text{total stock}}{\text{depth of cut}}$

Velocity of the workpiece V (m/min) $= \dfrac{\pi DN}{1000}$

SOLVED EXAMPLES

Example 1.1 : *Using a horizontal axis surface grinder, a flat surface of C65 steel of size 100 mm × 250 mm is to be ground. A grinding wheel with 250-mm diameter and 20-mm thickness is used. Calculate the grinder time required. Assume a table speed of 10 m/min and wheel speed of 20 m/s.*

Solution : The rpm of the grinding wheel

$$= \dfrac{1000 \times 20 \times 60}{\pi \times 250} = 1528 \text{ rpm}$$

Let approach distance = 125 mm

Time for one pass $= \dfrac{250 + 250}{10 \times 1000}$

$$= 0.05 \text{ minutes}$$

Assuming an in-feed rate of 5 mm/pass

Number of passes required = 100/5 = 20

Total grinding time $= 20 \times 0.05 = 1$ minute.

Fig. 1.21 shows the situation of a surface grinding operation using a vertical axis machine.

Approach distance for this case is given as,

$$A = \dfrac{D}{2} \text{ for } W = \dfrac{D}{2} \text{ up to } D$$

$$A = \sqrt{W (D - W)} \text{ for } W < \dfrac{D}{2}$$

where, $\qquad$ W = width of cut

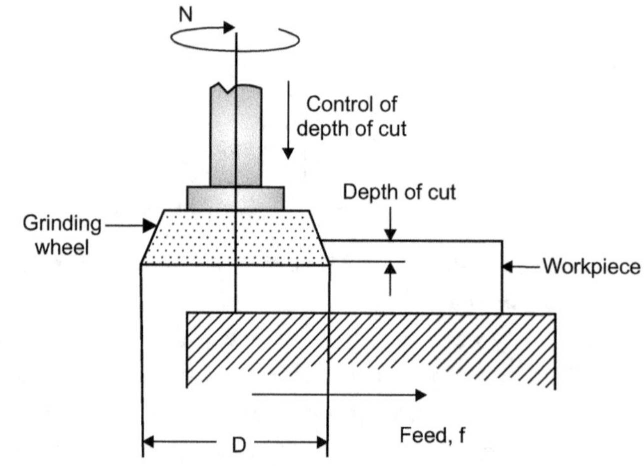

Fig. 1.21 : Grinding operation

Example 1.2 : *For the above example, if vertical axis surface grinder it to be used. Calculate the grinding time. The wheel to be used has a 200-mm diameter with a 20-mm wheel thickness.*

Solution : Given, W = 100 mm, and D = 200 mm

Approach distance, A = 100 mm

Total machining time $= \dfrac{250 + 200}{10 \times 1000} = 0.045$ minutes

Example 1.3 : *For rough grinding operations, determine the machining time required when cutting speed is 25m/min, Diameter of work is 45 mm. depth of cut is 0.03 mm, stock is 0.6 mm for 220 mm long work piece with face width of wheel is 70 mm.* $\qquad$ **[Feb. 15, 6M]**

Solution : We know, assume k = 1.2

Number of passes = n $= \dfrac{\text{stock}}{\text{depth of cut}} = \dfrac{0.6}{0.03} = 20$

Therefore,

Now, $\qquad$ V $= \dfrac{\pi DN}{1000}$

$$25 = \dfrac{\pi \times 45 \times N}{1000}$$

∴　Speed of wheel = N = 176.84 rpm

Now, Find longitudinal,

$$f = (0.7 \text{ to } 0.85) \times 70$$

$$0.76 \times 70 = 53.2 \cong 54 \text{ mm}$$

∴　Machining time = $T_m = \dfrac{L \times n \times k}{N \times f}$

$$= \dfrac{220 \times 20 \times 1.2}{54 \times 176.84} = 0.55 \text{ minutes.}$$

Example 1.4 : *For finish grinding operations, determine the machining time required when cutting speed is 20 m/min, Diameter of work is 40 mm, depth of cut is 0.03 mm, stock is 0.6 mm for 230 mm long work piece with face width of wheel is 70 mm (take k = 1.6 for finishing).*　　**[Feb. 16, 6M]**

Solution : L = 230 mm, D = 40 mm, depth of cut = 0.03 mm, B = 70 mm, k = 1.6 V = 20 m/min

∴　Number of passes (n) = $\dfrac{\text{total stock}}{\text{depth of cut}}$

$$= \dfrac{0.6}{0.03} = 20$$

Now,　$V = \dfrac{\pi \, DN}{1000}$

$$20 = \dfrac{\pi \times 40 \times N}{1000}$$

∴　　　N = 159.15 rpm

∴　For finish ginding,

$$f = (0.2 \text{ to } 0.4) \times b$$

$$= (0.2 \text{ to } 0.4) \times 70$$

$$f = 22 \text{ mm}$$

∴　Machining Time = $T_m = \dfrac{L \times n \times k}{N \times f} = \dfrac{230 \times 20 \times 1.6}{22 \times 159.15}$

$$T_m = 2.10 \text{ min}$$

Example 1.5 : *Determine the machining time required for rough grinding with following data..*

Cutting speed of 20 m/min, depth of cut 0.05 mm, the diameter of workpiece is 60 mm, length of workpiece 220 mm. The stock to be removed 0.35 mm. The width of grinding wheel is 60 mm.

Solution :　D = 60 mm

$$L = 220 \text{ mm}$$

$$W_g = 60 \text{ mm}$$

$$d = 0.05 \text{ mm}$$

$$v = 20 \text{ m/min}$$

Stock to be removed = 0.35 mm

Number pf passes (n) = $\dfrac{\text{Total stock to be removed}}{\text{depth of cost}}$

$$D = \dfrac{0.35}{0.05} = 7$$

Speed of grinding wheel (N)

$$V = \dfrac{\pi DN}{60}$$

$$\dfrac{20}{60} = \dfrac{\pi \times 60 \times 10^{-3} \times N}{60}$$

$$N = 106.103 \text{ rpm}$$

Feed (f) = (0.6 to 0.9) × width of grinding wheel

$$= (0.6 \text{ to } 0.9) \times W_g$$

$$= (0.6 \text{ to } 0.9) \times 60$$

$$= (24 \text{ to } 54) \text{ mm}$$

$$= 39 \text{ mm/rev}$$

$$k = 1 \text{ to } 1.2\ldots \text{ for rough grinding}$$

∴　　　K = 1

Machining time,

$$T = \dfrac{L \times n \times K}{f \times N}$$

$$= \dfrac{220 \times 7 \times 1}{39 \times 106.103}$$

$$T = 0.372 \text{ min}$$

$$T = 22.329 \text{ seconds}$$

Example 1.6: *Grinding operation is carried out at cutting speed of 35 m/min with depth of cut 0.03 mm. The diameter and length of the work is 55 mm and 250 mm respectively. The total stock to be removed is 0.6 mm. The width of grinding is 50 mm. Determine the machining time for finish grinding.*

Solution :　D = 55 mm

$$L = 250 \text{ mm}$$

$$W_g = 50 \text{ mm}$$

$$d = 0.03 \text{ mm}$$

$$V = 35 \text{ m/min}$$

Stock to be removed = 0.6 mm

Number of passes (n) = $\dfrac{\text{Total stock to be removed}}{\text{depth of cut}}$

$$= \dfrac{0.6}{0.03} = 20$$

Feed $= 0.6 \times$ face width of wheel

$$= 0.6 \times 50$$

$$= 30 \text{ mm/rev}$$

Speed of grinding Wheel (N),

$$V = \frac{\pi DN}{60}$$

$$\frac{35}{60} = \frac{\pi \times 55 \times 10^{-3} \times N}{60}$$

$$N = 202.56 \text{ rpm}$$

$$K = 1.3 \text{ to } 1.7 \ldots \text{ for finish grinding}$$

$$= 1.5$$

Machining time $T = \dfrac{L \times n \times K}{f \times N}$

$$= \frac{250 \times 20 \times 1.5}{30 \times 202.56}$$

$$T = 1.234 \text{ min}$$

$$= 74.052 \text{ seconds}$$

SUPER FINISHING PROCESSES

1.10　HONING　　　[Nov. 15; Feb., May 16]

1.10.1 Construction and Working

- Honing is an smooth process for removing material from metallic and non-metallic surfaces.

- The abrasive particles (Al_2O_3 or SiC) are held by proper bond in the form of stick are mounted on a mandrel which is then given a reciprocating movement (along the hole axis) superimposed on a uniform rotary motion.

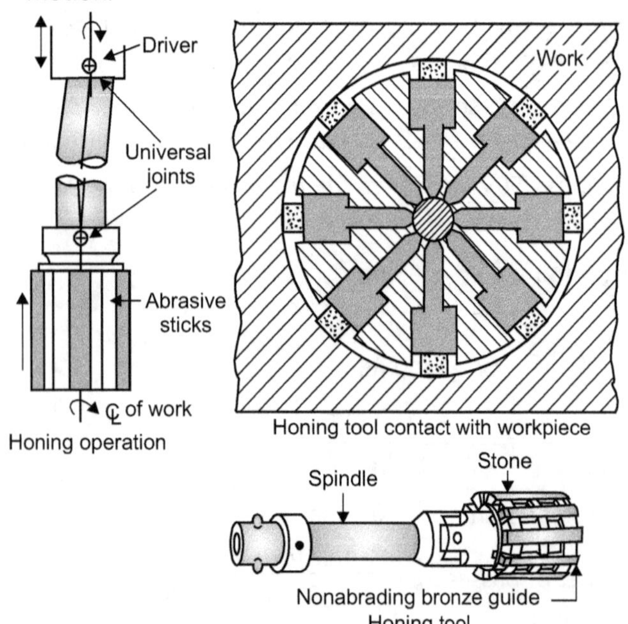

Fig. 1.22 : Honing

- Honing is employed for grinding internal or external surfaces. Mostly honing is done on internal surfaces or holes such as automobile cylinders.

- It is used to correct local irregularities such as waviness of axis, or non parallelism of cylindrical features and to develop a particular texture.

- Honing is the application of bonded abrasive stones to a surface for the purposes of limited stock removal and attainment of a surface finish.

- The stock removal is very low (0.5 mm for prime honing and 0.01 mm for secondary or mirror honing).

- It makes use of abrasive grit which shears off chip from the metal surface. As the edge becomes dull, pressure developed breaks the grit and new surface comes up and cutting action continues.

- Honing can correct some out of roundness, taper and axial distortion as it can remove upto 0.5 mm of stock.

- The honing operation is used for finishing the inside surface of a hole.

- The tolerance and finish achieved in this operation are of the order of 0.0025 mm and 0.25 μm respectively.

1.10.2 Controlling Parameters

- Material, practically any material can be honed, Soft materials, which cannot be lapped because the lapping particles get embedded into the metal can be honed because of the use of bonded abrasive.

- In any case, softer materials are difficult to be honed because of the loading they tend to cause. Metallic surfaces such as hard and soft cast iron, steel, carbides, brass, bronze, Al, Cr and also the glass, ceramics, plastics can be honed.

- Hardness, it does not affect the honing process but affects only the rate of stock removal.

- Size, smallest 6.25 mm and biggest may be any, limitation is set by the machine used.

- Shape, the work may be of an shape as long as surface being honed is cylindrical The bore may be interrupted by part, keyways and undercuts. Unusual condition may require special equipment but if there is enough area on the bore to stabilize the action of abrasive stick, it can be honed.

- If parts are being honed to correct the inaccuracies like out of roundness, there must be about twice the stock left for honing as there is error in the bore for finishing.

  ➢ Rough honing for stock removal.

  ➢ Finish honing for developing desired finish.

- To remove the marks of previous operation, 0.005 to 0.025 mm material should be left for honing.

- Surface Finish, honing gives characteristic cross-batched surface finish to surface. The depth of cut depends on the abrasive, speed, pressure, coolant and material. To get the fine finish, the following points should be remembered.

  ➢ Increase speed of rotation.

  ➢ Use fine grit.

  ➢ Increase viscosity of coolant.

  ➢ Decrease pressure.

1.11 LAPPING   [May, Nov. 15; Feb. 16]

1.11.1 Construction and Working

- Lapping is another operation for improving the accuracy and finish. It accomplished by abrasives in the range 120 – 1200 mesh. It is an abrading process for refining surface finish and the geometrical accuracy of flat cylindrical and spherical surfaces.

- A lap is generally made of a material softer than the work material.

- In this process, straight, narrow grooves are cut and this surface is charged by sprinkling the abrasive powder.

- The workpiece is then held against the lap and moved in unrepeated paths. A suitable cutting fluid is applied for lapping.

- The material removal is seldom more than 0.0025 mm and the lapping pressure is generally kept in the range 0.01 – 0.2 N/mm^2, depending on the hardness of the work material.

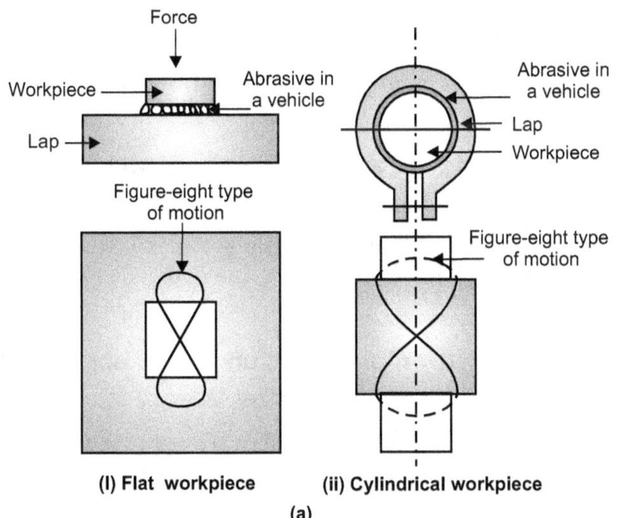

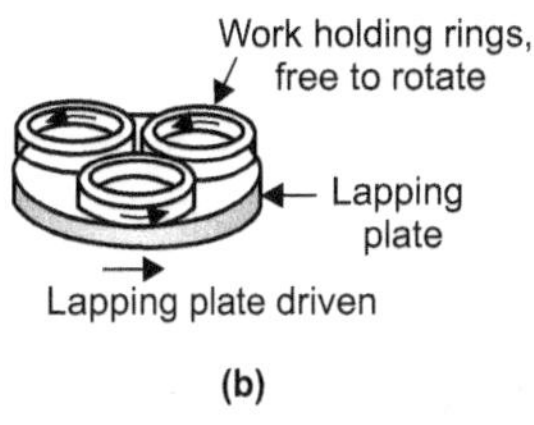

Fig. 1.23 : Lapping operation

- Lapping consists of the use of loose-grain abrasive flours mixed with oil.

- The abrasive is applied to the work by means of lapping shoes or quills which are made of softer metal than the work to be lapped.

- The lap is charged with a small amount of abrasive by spreading it over a hardened-steel plate and rolling or pressing the lap into the abrasive.

- The fine, sharp abrasive particles become embedded in the lap, and it is ready for use. In modern machine driven laps, a grease (mixture of abrasive and vehicle) is smeared on the face of lap which may be operated by hand or machine, the motion being rotary or reciprocating.

- Lap and work are not positively driven but are guided in contact with each other.

- The fresh contacts are made between lap and work through constantly changing relative movements most effective when cycloidal in nature.

- Types of lapping operations are equalising lapping, when work and lap mutually improve their shape and surface, for example ; when gears are run together with some abrasive, or tapered valves are seated in seats.

- Form lapping, shape of lap is imparted to work.

1.11.2 Controlling Parameters

- **Lap Material**, cast iron is mostly used for machine lapping. The lap material should be soft (softer than the workpiece so that the abrasive compounds get embedded in the lap), close grained, free from porosity and surface defects.

- When cast iron is not suitable, steel, brass, copper, type metal, aluminium may be used. For lapping by diamond, diamond is hammered into the metal where it is embedded permanently. Soft laps are used for special purpose and for super finishing. Wood is also sometimes used for certain applications.

- **Abrasive**, hard abrasives are used for harder work materials and soft abrasives for soft work materials; Diamond is hardest material and is used for lapping

WC and precious stones. Boron carbide, silicon carbide and Al_2O_3. Grain sizes run from 60 to 100 grit, the latter is a very fine powder. **SiC** is used for rapid stock removal, and Al_2O_3 for improved surface.

- **Vehicle,** abrasives are mixed with carrier medium called vehicle. Its purpose is to suspend abrasive and keep grains separated as well as to lubricate the work and prevent scoring.

- **Speed and Pressure,** speeds of the order of 1.5 to 4 m/sec are used when relative motion between lap and work is rotary. Pressures of the order of 0.07 to 0.2 kg/cm^2 are used for soft materials and 0.7 kg/cm^2 for hard materials. Pressure greater than this would result in it breakdown and scoring of workpiece.

- Materials to be lapped, softer and non-ferrous materials require finer grit size to obtain good finish.

- **Lapping Process,** for hand lapping, a plate is used on which work is rubbed with abrasive on it, the rubbing motion is a figure of eight (8) to cover entire surface of plate. For lapping holes laps should be such as to range from diameter of unlapped hole to lapped finished hole.

1.12 BUFFING

1.12.1 Construction and Working

- It is a polishing operation in which the component is in contact with a rotating cloth buffing wheel that usually has been charged with a very fine abrasive Fig. 1.21.

- The polishing action in buffing is similar to lapping in that when a polishing medium such as 'rouge' is used; the cloth buffing wheel becomes a carrying vehicle for the fine abrasives.

- It eliminating the scratch marks and producing a very smooth surface. When softer metals are buffed, mainly without the use of an abrasive, there is some indication that a small amount of metal flow may occur which helps to reduce the high spots and produce a high polish.

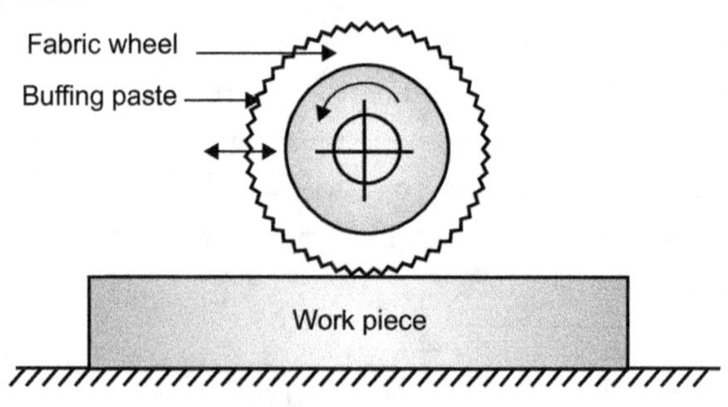

Fig. 1.24 : Buffing

- Buffing wheels are commonly made of discs of linen, cotton, broad cloth and canvass materials.

- They are made more or less definite by the amount of stitching used to fasten the layers of the cloth together.

- Buffing wheels for very soft polishing or which can be used to polish into interior corners may have no stitching, the cloth layers being kept in position by the centrifugal force resulting from the rotational movement of the wheel.

- Various types of buffing rouges are available. Most of them being mainly ferric oxide in some soft type of binder.

- Buffing should be used only to remove very fine scratches or to remove oxide or similar coatings which may be on the work surface.

- There are semi-automatic buffing machines available consisting of a series of individually driven buffing wheel which can be adjusted to the desired position so as to buff different portions of the component.

- The component is held in fixtures on a rotating circular worktable so as to move past the buffing wheels.

- Applications of buffing process which produces mirror-like finish are objects used on automobiles, motor-cycles, mobile homes, boats, bicycles, sporting items, tools, store fixtures, commercial and residential hardware and household utensils and appliances.

1.12.2 Controlling Parameters

- **Satin Finishing,** it is done with a arranged compound of glue and binder applied to buff. The purpose is to polish or to produce matte or scratch decorative finish.

- Buffing wheel speeds are in the range of 32.5 to 40 m/s.

- **Setup Wheels,** are wheels to which abrasive grains are attached by coating them with a bond and rolling in the abrasive grains.

- **Paste Head Wheel,** a wheel coated with bond and grain formerly mixed; fine grains are often applied as a paste. Bond and grains (generally used is Al_2O_3) are also applied by spraying on the wheel when it is in motion.

- **Abrasive Belt-polishing,** factory coated abrasive belts coated with grains of all sizes are made. These are operated over contact wheels typically of buff type. Back stand a separate device to controlled tension and tracking of the belt.

1.13 BURNISHING [Nov. 16, 17; Feb. 17, May 17]

1.13.1 Construction and Working

- Burnishing operation is the process of getting a smooth and shiny surface by contact and rubbing of the surface against the surface of a hard burnishing tool.

- It is a finishing and strengthening process, basically a cold surface plastic deformation process.

- Cold working of improves the surface finish and induces surface compressive residual stresses, thus improving the fatigue life of the component.

- The sizing operation perform by a punch or mandrel and a die to get burnished inside and surfaces, inside surface of the bush is burnished when the mandrel gets forced through it the external surface is burnished when the bush is forced through the die.

- A special hardened gear shaped burnishing die subjects the tooth surfaces to a surface rolling action, the gear is rolled under press hardened accurately formed burnishing gear.

- It is obtained in any high points on the tooth surface are plastically deformed to get accurate and finished tooth profile, in cold working process.

1. Barrel Burnishing

- It gives approximately some output those obtained by buffing may be obtained by barrel burnishing.

- It is similar to barrel rolling except that instead of using abrasive medium, medium balls, shots or round pins are added to the work in the barrel.

- No cutting action in burnishing, it will not ordinarily remove visible scratches or pits, but will produce a smooth, uniform surface and reduce the porosity in surfaces which are to be or have been plated.

- First of all parts should be rolled with a fine abrasive, before barrel burnishing.

- Barrel burnishing normally is done wet condition, using mixture of water and some lubricating or cleaning agents such as soap.

- The barrel should not be loaded more than half full with work and shot. Since the rubbing action between the work and the shot material is very important, there should be about two volumes of shots to one volume of parts.

- The ratio should be maintain like, the workpiece do not rub against one another.

- The speed of the barrel should be adjusted so that the workpiece is not thrown out of the mass as they reach the top position and roll down the inclined surface.

2. Roller/Ball Burnishing

- Both internal and external flat, cylindrical or conical surfaces are burnished with hardened steel or cemented carbide rollers or with steel balls mounted in a holder.

- Fillets and grooves are burnished by rollers rounded to a radius.

- The aim of the treatment is strengthening, the burnishing pressure is to be increased, and this condition results in somewhat lower machining accuracy.

- Internal surfaces are also burnished with the help of balls, the process being called as **"Ballizing" or "Ball Burnishing"**.

- Smooth balls or mandrels slightly larger than the bore diameter are pushed through the length of the hole (Fig. 1.25).

- Common applications of roller burnishing include : Hydraulic system components, seals, valves, spindles and fillets on shafts.

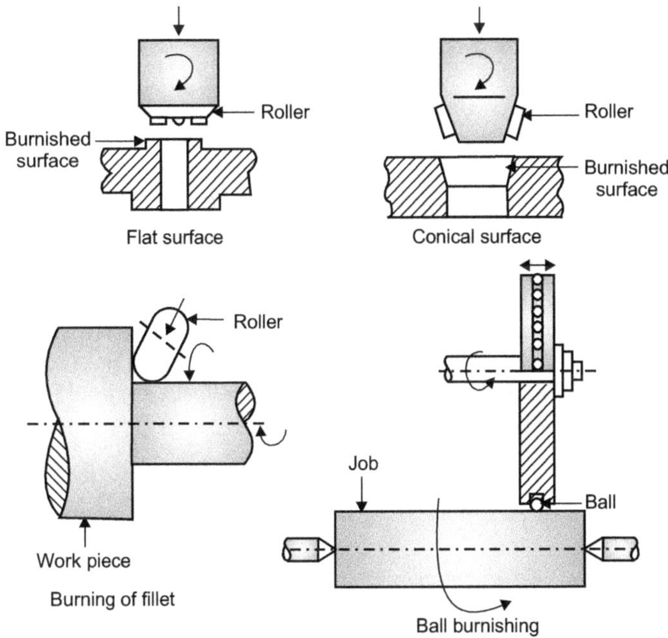

Fig. 1.25 : Roll burnishing

1.13.2 Controlling Parameters

- In barrel burnishing usually necessary to use several sizes and shapes of shot material in order to ensure that the material can come in contact with inside corners and other recesses which must be rubbed.

- Balls from 3 mm to 6 mm diameter, pins, jacks and ball cones are commonly used.

- Parts which cannot be permitted to bump against one another may be burnished successfully by fastening them in racks inside the barrel.

- The shot material is then added and burnishing carried out in the common manner. When proper conditions have been achieved, barrel burnishing is economical and produces surfaces suitable for subsequent painting or plating.

- Hole burnishing is performed with multi roller tools on drill presses, turret lathes, horizontal borers, unit built machines and automatic lathes.

- Ball Burnishing raises the hardness of the surface by 20 to 50% and its wear resistance by 1.5 to 2 times.

EXERCISE

1. What is the difference between rough grinding and precision grinding? Why is grinding so important in modern production?

2. Why the natural abrasives are not suitable for making grinding wheels?

3. How aluminium oxide is abrasive produced? Write its field of application.

4. How silicon carbide is abrasive produced? Write its field of application.

5. Under what conditions are diamond, boron carbide and cubic boron nitride used as abrasive materials for making grinding wheels?

6. What is meant by 'grain size' of an abrasive material?

7. Discuss the various types of bonding materials used for making grinding wheels?

8. Define a grinding wheel. Discuss the various methods of making grinding wheels.

9. Sketch the various shapes of grinding wheels and write their fields of application.

10. What is meant by "grade" and "structure" of a grinding wheel?

11. What is meant by standard marking of grinding wheels?

12. How the grinding wheel is selected for a particular job?

13. What is meant by dressing and truing of grinding wheels?

14. Define "grinding ratio".

15. Sketch and explain the working of an external cylindrical grinding machine.

16. Sketch and explain the three methods of cylindrical grinding.

17. What is the difference between plain and universal cylindrical grinders?

18. Describe in detail how an internal grinder operates.

19. What are the advantages and disadvantages of centreless grinding ?

20. Sketch and explain the three methods of external cylindrical centreless grinding.

21. Sketch and explain internal Centre less grinding process.

22. What are the various abrasive machining operations you are familiar with? Explain their application and limitations.

23. How is grinding different from other machining operations? Explain its applications in view of its capabilities.

24. What is the classification method that could be used for the grinding machine? Give the applications of each of the variety of grinding machine.

25. How is the abrasive selected for a grinding operation? Your answer should indicate the reasons for the selection.

26. What is the marking system followed in case of grinding wheels? Explain the individual elements of the marking system from the standpoint of the functioning of the wheel.

27. What are the grinding process parameters that are of interest? Explain their effect on the grinding performance and the wear rates.

28. What are the advantages and limitations of using centre-less grinding?

◈ ◈ ◈

MECHANICS OF METAL CUTTING

2.1 INTRODUCTION

- Metal cutting are also known as conventional machining processes.

- These processes are commonly carried out in machine shops or tool room for machining cylindrical or flat jobs to a desired shape, size and finish on a rough block of job material with the help of a wedge shaped tool.

- A machine tool is a power driven metal cutting machine which assist in managing the needed relative motion between cutting tool and the job that changes the size and shape of the job material.

- The machine tools involve various kinds of machines tools commonly named as lathe, shaper, planer, slotter, drilling, milling and grinding machines etc.

- The machining jobs are mainly of two types namely cylindrical and flats. Cylindrical jobs are generally machined using lathe, milling, drilling and cylindrical grinding whereas flat jobs are machined using shaper, planner, milling, drilling and surface grinding.

2.1.1 Purpose of Machining

- Most of the engineering components such as gears, bearings, clutches, tools, screws and nuts etc. need dimensional and form accuracy and good surface finish for serving their purposes.

- Manufacturing processes such as casting, forging etc. generally cannot provide the desired accuracy and finish.

- For that such preformed parts, called blanks, need semi-finishing and finishing and it is done by machining and grinding. Grinding is also basically a machining process.

- Machining to high accuracy and finish essentially enables a product

  ➢ Fulfill its functional requirements

  ➢ Improve its performance

  ➢ Prolong its service

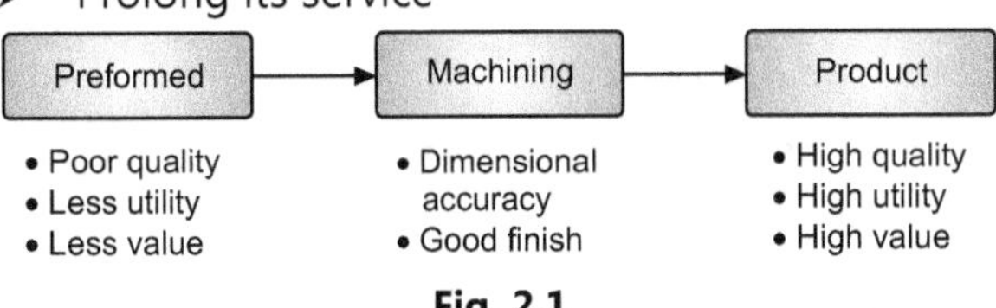

Fig. 2.1

2.1.2 Principle of Machining

The basic principle of machining is typically illustrated in Fig. 2.2.

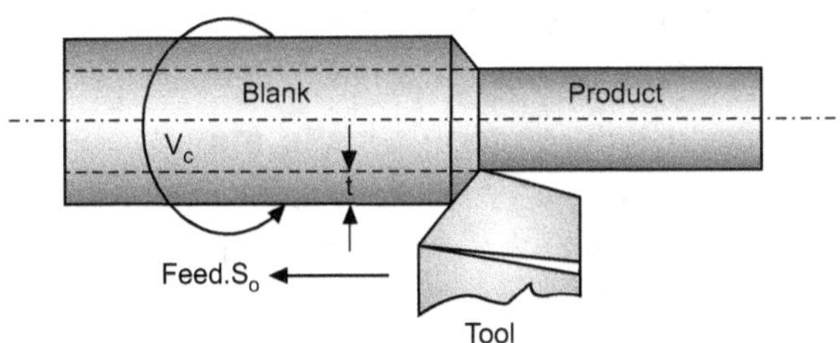

Fig. 2.2 : Principle of machining (turning)

A metal rod of irregular shape, size and surface is converted into a finished rod of desired dimension and surface by machining by proper relative motions of the tool-work pair.

2.1.3 Definition of Machining

- Machining is an essential process of finishing by which jobs are produced to the desired dimensions and surface finish by gradually removing the excess material from the preformed blank in the form of chips with the help of cutting tool(s) moved past the work surface(s).

- The essential basic requirements for machining work are schematically illustrated in Fig. 2.3.

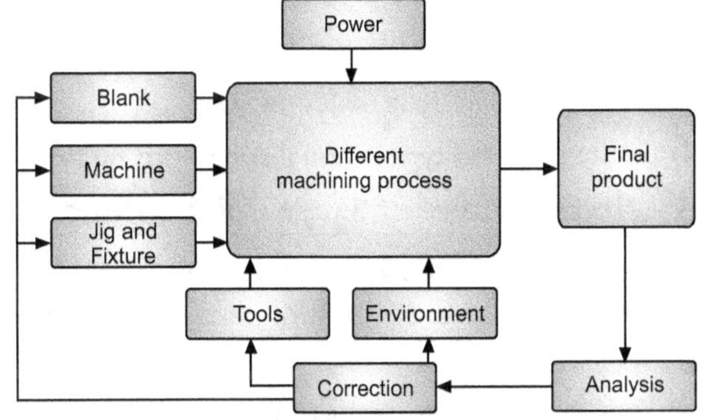

Fig. 2.3 : Requirements for machining

2.2 CUTTING TOOL

- Cutting tools performs the main machining operation. A single point cutting tool (such as a lathe, shaper and planner and boring tool) has only one cutting edge, whereas a multi-point cutting tool (such as milling cutter, milling cutter, drill, reamer and broach) has a number of teeth or cutting edges on its periphery.

2.3 SINGLE POINT CUTTING TOOL

[May, Nov. 15; Feb. 17]

- There are mainly two types of single point tools namely the solid type as shown in Fig. 2.4 (a) and the tipped tool Fig. 2.4 (b). The solid type single point tool may be made from high speed steel, from a cast alloy.

- Brazed tools are generally known as tool bits and are used in tool holders.

- The tipped type of tool is made from a good shank steel on which is mounted a tip of cutting tool material. Tip may be made of high speed steel or cemented carbide.

- The insert type Fig. 2.4 (c) tool throw away refers to the cutting tool insert which is mechanically held in the tool holder. The inserts are purchased which are ready for use. Geometry comprises mainly of nose, rake face of the tool, flank, heel and shank etc. The nose is shaped as conical with different angles.

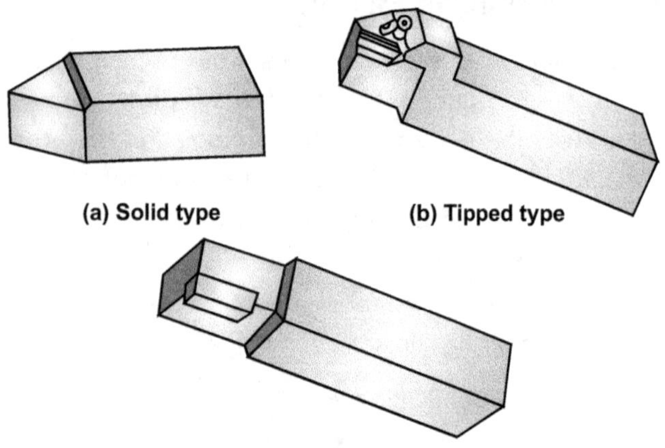

(a) Solid type   **(b) Tipped type**

(c) Index - able insert type

Fig. 2.4 : Different types of single point cutting tool

2.3.1 Nomenclature of Single Point Cutting Tool

Single point cutting tool is widely used in machining operations. The detail description of parts and terminology as follows

- **Shank :** It is the body of the tool which is ungrounded.

- **Face :** It is the surface over which the chip slides.

- **Base :** It is the bottom surface of the shank.

- **Flank :** It is the surface of the tool facing the work piece. There are two flanks namely end flank and side flank.

- **Cutting Edge :** It is the junction of the face and the flanks. There are two cutting edges namely side cutting edge and end cutting edge.

- **Nose :** It is the junction of side and end cutting edges.

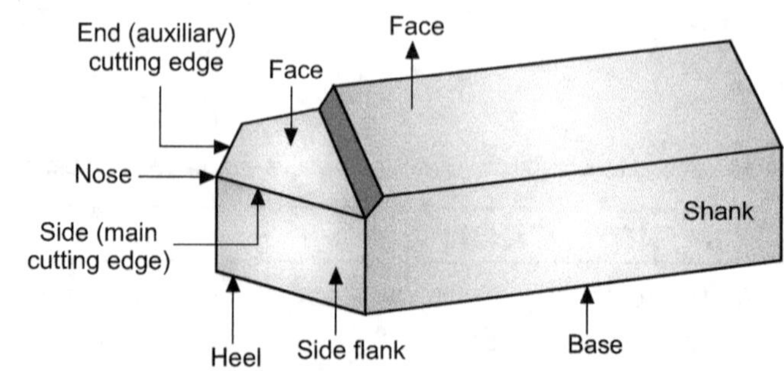

Fig. 2.5 : Geometry of single point cutting tool

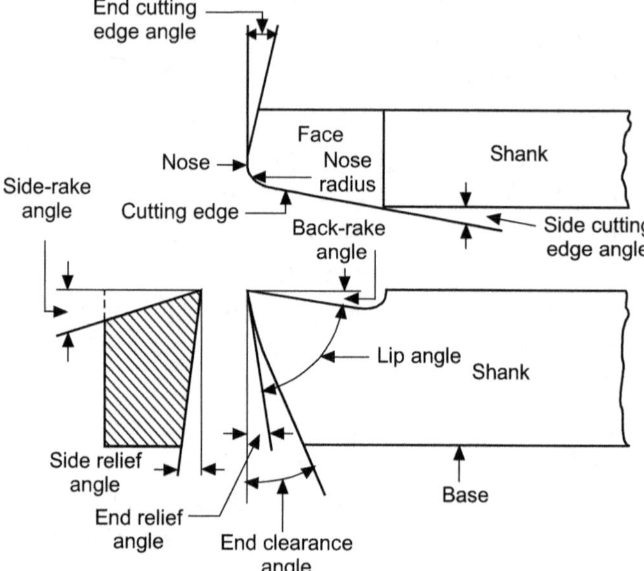

Fig. 2.6 : Nomenclature of single point cutting tool

Different angles are playing major role in machining. Following are important angles.

1. **Back Rake Angle**

   - It is the angle between the face of the tool and a line parallel with base of the tool measured in a perpendicular plane through the side cutting edge.

   - This angle helps in removing the chips away from the work piece.

   - The rake angle is always at the topside of the tool.

   - It influence on tool strength. A tool with negative rake will withstand far more loading than a tool with positive rake.

   - It influence on cutting pressure. A tool with a positive rake angle reduces cutting forces by allowing the chips to flow more freely across the rake surface.

2. **Side Rake Angle**

   - It is the angle by which the face of tool is inclined sideways.

   - This angle of tool determines the thickness of the tool behind the cutting edge.

3. Relief Angle

- It is the angle between the tool flank and a normal line drawn from the cutting point to the tool base
- Relief angles minimize physical interference or rubbing contact with machined surface and the work piece.
- Relief angles eliminate tool breakage and to increase tool life.

(i) Side Relief Angle

➢ The Side relief angle prevents the side flank of the tool from rubbing against the work when longitudinal feed is given.

➢ Larger feed will require greater side relief angle.

(ii) End Relief Angle

➢ The End relief angle prevents the end flank of the tool from rubbing against the work.

➢ A minimum relief angle is given to provide maximum support to the tool cutting edge by increasing the lip angle.

4. End Cutting Edge Angle

- It is the angle between the end cutting edge and a line perpendicular to the shank of the tool.

5. Side Cutting Edge Angle

- It is the angle between straight cutting edge on the side of tool and the side of the shank.

6. Nose Radius

- It is the nose point connecting the side cutting edge and end cutting edge.
- It possesses small radius which is responsible for generating surface finish on the work-piece.

2.4 METHODS OF MACHINING

In metal cutting operation, the position of cutting edge of the cutting tool is important based on which the cutting operation is classified as orthogonal and oblique cutting.

- **Orthogonal Cutting (2D)** is also known as two dimensional metal cutting in which the cutting edge is normal to the work piece. In orthogonal cutting no force exists in direction perpendicular to relative motion between tool and work piece.

- **Oblique Cutting (3D)** is the common type of three dimensional cutting used in various metal cutting operations in which the cutting action is inclined with the job by a certain angle called the inclination angle.

2.4.1 Comparison between Orthogonal Cutting (2D) and Oblique Cutting (3D)

Sr. No.	Orthogonal Cutting (2D)	Oblique Cutting (3D)
1.	Fig. 2.7	Fig. 2.8
2.	The cutting edge of the tool remains normal to the direction of tool feed or work feed.	The cutting edge of the tool remains inclined at an acute angle to the direction of tool feed or work feed.
3.	The direction of the chip flow velocity is normal to the cutting edge of the tool.	The direction of the chip flow velocity is at an angle with the normal to the cutting edge of the tool. The angle is known as chip flow angle.
4.	Here only two components of forces are acting : Cutting Force and Thrust Force. So the metal cutting may be considered as a two dimensional cutting.	Here three components of forces are acting : Cutting Force, Radial force and Thrust Force or Feed Force. So the metal cutting may be considered as a three dimensional cutting.
5.	The chip curls and flows straight up the tool and not side ways.	The chip flows side ways
6.	The width of the tool is more.	It may or may not.
7.	Heat developed per unit area due to friction along the tool-w/p interface is considerably more.	Heat developed per unit area is less.
8.	Cutting Angle = 90	Cutting Angle < 90
9.	E.g. Jack plane operation in carpentry. Parting off operation in lathe, Sawing operation	E.g. Turning, milling, drilling, shaping, grinding etc.

2.5 TOOL SIGNATURE/DESIGNATION

- Convenient way to specify tool angles by use of a standardized abbreviated system is known as tool signature or tool nomenclature.

- The signature is the sequence of numbers listing the various angles, in degrees, and the size of the nose radius.

- There are several systems available like American standard Association system (ASA), orthogonal rake system (ORS), Normal rake system (NRS), and Maximum rake system (MRS).

- The system most commonly used is American Standard Association (ASA).

2.5.1 American Standard Association (ASA)

- In ASA system, the angle of tool face that, is its slope, are defined in two orthogonal planes, one parallel and other perpendicular to the axis of cutting tool, both plane being perpendicular to the base of tool.

- The seven elements that comprise the signature of a single point cutting tool can be stated in the following order :

  Back rake angle, Side rake angle, End relief angle, Side relief angle, End cutting Edge angle, Side cutting Edge angle and Nose radius.

Tool Signature : $\alpha_b - \alpha_s - \theta_e - \theta_s - C_e - C_s - R$

1. Back rake angle (α_b)
2. Side rake angle (α_s)
3. End relief angle (θ_e)
4. Side relief angle (θ_s)
5. End cutting edge angle (C_e)
6. Side cutting edge angle (C_s)
7. Nose radius (R)

For Example:

Tool Signature: $0 - 7 - 6 - 8 - 15 - 16 - 0.8$mm

1. Back rake angle (0°)
2. Side rake angle (7°)
3. End relief angle (6°)
4. Side relief angle (8°)
5. End cutting edge angle (15°)
6. Side cutting edge angle (16°)
7. Nose radius (0.8 mm)

- In ASA system of tool angels, the angels are specified independently of the position of the cutting edge. But in actual cutting operation, we should include the side cutting edge (principle cutting edge). Such a system is known as orthogonal rake system (ORS).

2.5.2 Orthogonal Rake System (ORS)

The tool designation under ORS is,

Inclination angle, Rake angle, Side relief angle, End relief angle, End cutting edge angle, Approach angle, Nose radius.

Tool Signature : $i - \alpha - \gamma - \gamma i - C_e - \lambda - R$

1. Inclination angle (i)
2. Rake angle (α)
3. Side relief angle (γ)
4. End relief angle (γ_i)
5. End cutting edge angle (C_e)
6. Approach angle (λ)
7. Nose radius (R)

For Example :

Tool Signature : $0 - 10 - 6 - 6 - 8 - 10 - 1$ mm

1. Back rake angle (0°)
2. Side rake angle (10°)
3. End relief angle (6°)
4. Side relief angle (6°)
5. End cutting edge angle (8°)
6. Side cutting edge angle (10°)
7. Nose radius (0.8 mm)

2.5.3 Relation between ASA System and ORS

$$\tan \alpha = \tan \alpha_s . \sin \lambda + \tan \alpha_b . \cos \lambda$$
$$\tan \alpha_b = \tan \alpha . \cos \lambda + \tan i . \sin \lambda$$
$$\tan \alpha_s = \tan \alpha . \sin \lambda - \tan i . \cos \lambda$$
$$\tan i = - \tan \alpha_s . \cos \lambda + \tan \alpha_b . \sin \lambda$$

2.6 MECHANICS OF METAL CUTTING

- The work piece is securely clamped in a machine tool vice or clamps or chuck.

- A wedge shape tool is set to a certain depth of cut and is forced to move in direction as shown in Fig. 2.9.

- All traditional machining processes require a cutting tool having a basic wedge shape at the cutting edge.

- The tool will cut or shear off the metal, provided

  ➤ The tool is harder than the metal.

  ➤ The tool is properly shaped so that its edge can be effective in cutting the metal.

  ➤ The tool is strong enough to resist cutting pressures but keen enough to sever the metal.

> Provided there is movement of tool relative to the material or vice versa, so as to make cutting action possible.

- When the tool advances into the work piece, the metal in front of the tool is severely stressed.

- The cutting tool produces internal shearing action in the metal. The metal below the cutting edge yields and flows plastically in the form of chip.

- Compression of the metal under the tool takes place.

- When the ultimate stress of the metal is exceeded, separation of metal takes place.

- The plastic flow takes place in a localized area called as shear plane.

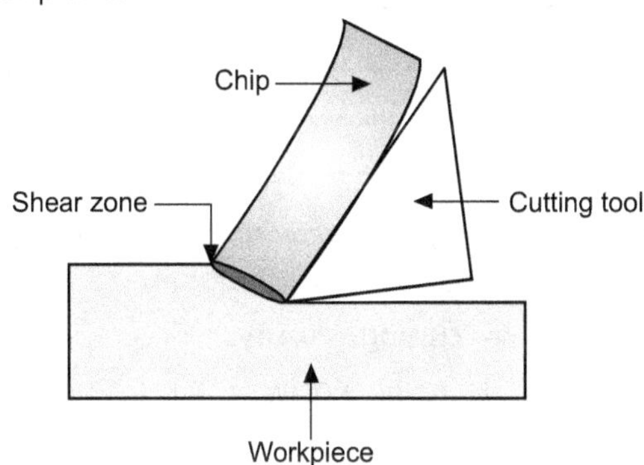

Fig. 2.9 : Mechanics of metal cutting

2.7 TYPES OF CHIPS [Nov. 15, Feb. 16, 17, May 16]

The chips that are formed during metal cutting operations can be classified mainly into three types :

1. Continuous chips
2. Discontinuous or segmental chips
3. Continuous chips with built-up edge.

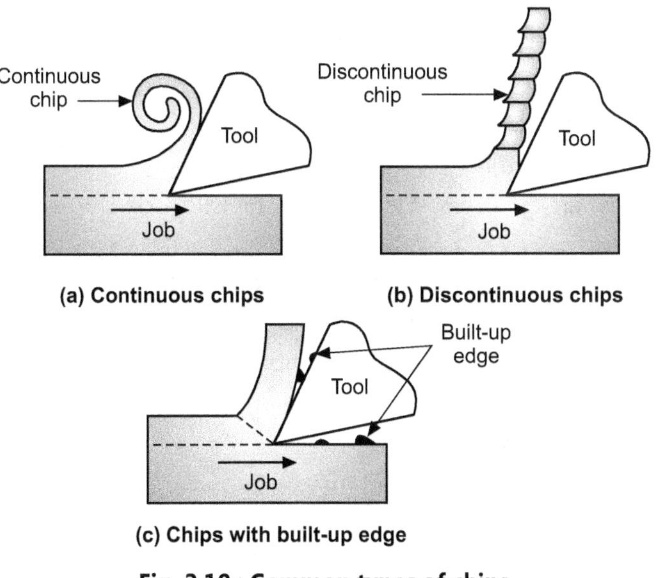

Fig. 2.10 : Common types of chips

1. **Continuous Chips**
   - Such chips are produced while machining ductile materials like mild steel, copper and aluminum.
   - Because of plastic deformation of ductile material long and continuous chips are produced.
   - The conditions for continuous chips,
     - Small chip thickness,
     - High cutting speed,
     - Sharp cutting edge,
     - Large rake angle in cutting tool and fine feed,
     - Smooth tool face.
   - This is desirable because it produces good surface finish, low power consumption and longer tool life.
   - These chips are difficult to handle and dispose off. Further the chips coil in a helix and curl around work and tool and may injure the operator when it is breaking.
   - To avoid this chip breakers are used. Chip breaker can be an integral part of the tool design or a separate device.

2. **Discontinuous or Segmental Chips**
   - In this type, the chip is produced in the form of small pieces.
   - These chips are produced when cutting more brittle materials like bronze, hard brass and gray cast iron.
   - Fairly good surface finish is obtained and tool life is increased with this type of chips.
   - Discontinuous chips are produced in ductile materials under the conditions such as
     - Large chip thickness,
     - Low cutting speed,
     - Small rake angle of tool,
     - Cutting fluids etc.
   - If these chips are produced from brittle materials, then the surface finish is fair, power consumption is low and tool life is reasonable.
   - These are convenient to handle and dispose off.

3. **Continuous Chip with Built-up Edge**
   - During cutting operation, the temperature rises and as the hot chip passes over s the face of the tool, alloying and welding action may take place, which results in the formation built-up edge.
   - This is nothing but a small built up edge sticking to the nose/face of the cutting tool. These built up edge occurs with continuous chips.

- When machining ductile materials due to conditions of high local temperature and extreme pressure the cutting zone and also high friction in the tool chip interface, there are possibilities of work material to weld to the cutting edge of tool and thus forming built up edges.

- This weld metal is extremely hard and brittle. This welding may affect the cutting action of tool.

- Successive layers are added to the build-up edge. When this edge becomes large and unstable it is broken and part of it is carried up the face of the tool along with chip while remaining is left in the surface being machined.

- Low cutting speeds lead to the formation of built up edge, however with high cutting speeds associated with sintered carbide tools, the build up edge is negligible or does not exist.

- Conditions favoring the formation of build up edge are

  - Low cutting speed,
  - Low rake angle,
  - High feed and large depth of cut.

- This formation can be avoided by the use of coolants and taking light cuts at high speeds. This leads to the formation of crater on the surface of the tool.

2.8 CHIP BREAKER   [Nov. 16, 17; May 17]

- Continuous chips are difficult to handle and dispose off. Further the chips coil in a helix and curl around work and tool and may injure the operator when it is breaking.

- Chip breaker is a piece of metal clamped to the rake surface of the tool which bends the chip and breaks it.

- Chips can also be broken by changing the tool geometry, thereby controlling the chip flow.

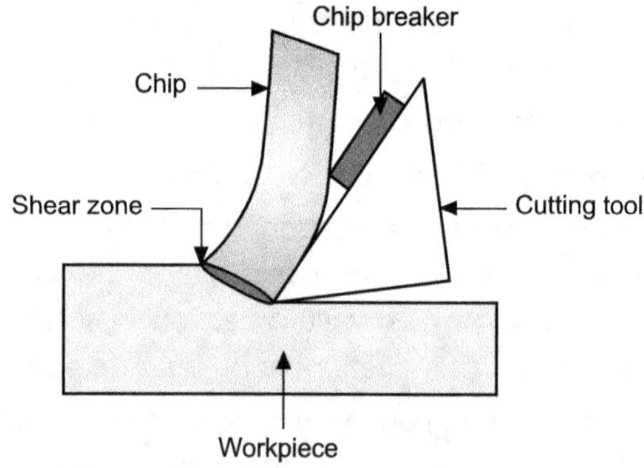

Fig. 2.11 : Chip breaker

2.9 ANALYSIS OF ORTHOGONAL CUTTING WITH SINGLE POINT CUTTING TOOL / CHIP-THICKNESS RATIO AND SHEAR ANGLE   [Feb. 15; May 16]

The outward flow of the metal causes the chip to thicker after separation from the parent metal. Metal prior to being cut is much longer than the chip which is removed.

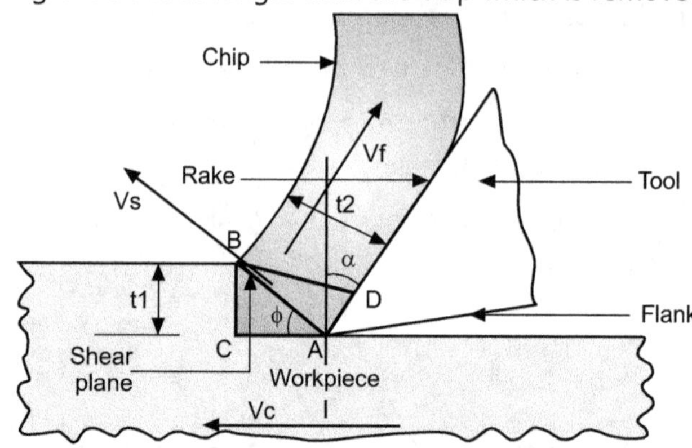

Fig. 2.12 : Analysis of orthogonal cutting with single point cutting tool

where, V_c = cutting velocity

      V_s = shear velocity

      V_f = velocity of chip-flow up the tool face.

      t_1 = chip thickness before cutting

      t_2 = chip thickness after cutting

      ϕ = shear angle

      α = rake angle

Chip thickness ratio,

$$r = \frac{t_1}{t_2}$$

- The chip thickness ratio is always less than unity.

$$K = \text{chip reduction co-efficient} = \frac{1}{r}$$

- When metal is cut there is no change in the volume of the metal cut.

$$t_1 b_1 l_1 = t_2 b_2 l_2$$

where, b_1 = Width of cut or width of chip before cutting

      l_1 = length of chip before cutting

      b_2 = width of chip after cutting

      l_2 = length of chip after cutting

Material is incompressible i.e. Volume before cutting same as volume after cutting.

i.e. $l_1 b_1 t_1 = l_2 b_2 t_2$

It is observed that,

$$b_1 = b_2$$
$$t_1 l_1 = t_2 l_2$$

Chip thickness ratio can be obtained by measuring l_1 and l_2

From $\triangle$ ACB, we have

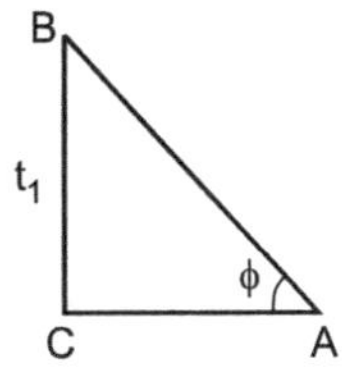

Fig. 2.13

$$\frac{BC}{AB} = \sin \phi$$

$$AB = \frac{BC}{\sin \phi} = \frac{t_1}{\sin \phi} \qquad \dots (2.1)$$

Let us consider $\triangle$ ABD,

From $\triangle$ ADB

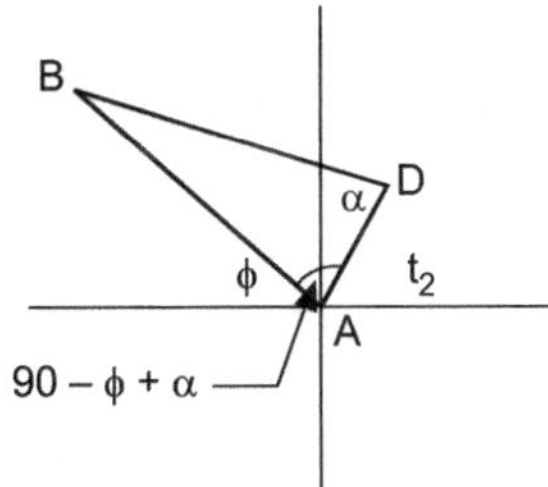

Fig. 2.14

$$\frac{BD}{AB} = \sin (90 - \phi + \alpha) = \cos (\phi - \alpha)$$

$$\frac{t_2}{AB} = \cos (\phi - \alpha)$$

$$AB = \frac{t_2}{\cos (\phi - \alpha)} \qquad \dots (2.2)$$

From equation (2.1) and (2.2), we get

$$\frac{t_1}{\sin \phi} = \frac{t_2}{soc (\phi - \alpha)}$$

$$\frac{t_1}{t_2} = \frac{\sin \phi}{\cos (\phi - \alpha)} = r$$

$$r = \frac{\sin \phi}{\cos \phi \cos \alpha + \sin \phi \sin \alpha}$$

Divide by $\cos \phi$ in numerator as well as denominator,

$$r = \frac{\sin \phi / \cos \phi}{(\cos \phi \cos \alpha + \sin \phi \sin \alpha)/ \cos \phi}$$

$$r = \frac{\tan \phi}{(\cos \alpha + \tan \phi \sin \alpha)}$$

$$r \cos \alpha + r \tan \phi \sin \alpha = \tan \phi$$

$$r \cos \alpha = (1 - r \sin \alpha) \tan \phi$$

$$\therefore \quad \tan \phi = \frac{r \cos \alpha}{1 - r \sin \alpha}$$

The closer the shear angle approaches 45°, the better the machinability is said to be.

2.10 VELOCITY RELATIONSHIPS

[Nov. 16, 17; May 17]

There are mainly three types of velocities involved in any metal cutting operations.

where, V_c = cutting velocity

V_s = shear velocity

V_f = velocity of chip-flow up the tool face

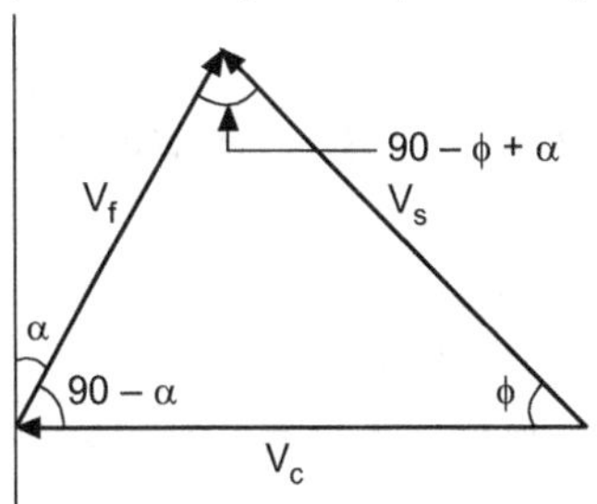

Fig. 2.15 : Velocity triangle

By applying sine principle,

$$\frac{V_f}{\sin \phi} = \frac{V_c}{\sin (90 - \phi + \alpha)} = \frac{V_s}{\sin (90 - \alpha)}$$

$$\frac{V_f}{\sin \phi} = \frac{V_c}{\cos (\phi - \alpha)} = \frac{V_s}{\cos (\alpha)}$$

By volume flows rate principle,

Volume before cutting = volume after cutting

Then,

$$t_1 b_1 V_c = t_2 b_2 V_f \text{ [width remains constant based on assumption]}$$

$$\frac{V_f}{V_c} = \frac{t_1}{t_2} = r$$

For strain rate calculation consider following Fig. 2.16.

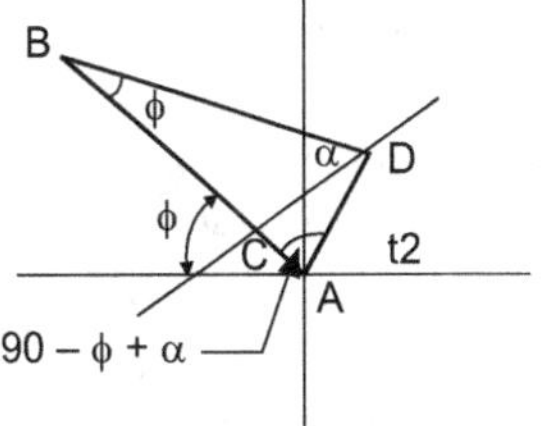

Fig. 2.16 : Shear zone

$$\therefore \quad \text{Shear strain} = \frac{\text{distance sheared}}{\text{thickness of shear zone}}$$

$$s = \frac{AB}{DC} = \frac{(AC + CB)}{DC}$$

$$= \left(\frac{AC}{DC}\right) + \left(\frac{CB}{DC}\right)$$

$$s = \cot\phi + \tan(\phi - \alpha)$$

To produce minimum shear strain in chip difference shear strain w.r.t. to ϕ and equate to zero.

Optimum or minimum shear angle is $2\phi - \alpha = 90$

$\therefore$ Shear strain rate $= \dfrac{V_s}{\text{thickness of shear zone}}$

$$\dot{s} = \dfrac{V_s}{t_s}$$

where, t_s is approx. 2.5×10^{-3} mm

From principle of kinematics that the relative velocity of two bodies (tool and chip) is equal to the vector difference between their velocities relative to the reference body.

$$\vec{V_c} = \vec{V_s} + \vec{V_f}$$

2.11 MERCHANT'S CIRCLE OF FORCES

[Nov. 15; May 16]

Following Forces are acting during machining operation,

- F – Frictional force at chip tool interface.
- N – Force normal to frictional force.
- F_s – Shear force along share plane.
- F_n – Force normal to shear plane.

These four actual forces acting in machining are the dynamic environment therefore it is difficult to measure them.

Hence each and every force can be resolved into two components i.e. horizontal and vertical.

Let,

F_c – Algebraic sum of vertically resolved forces (cutting force)

F_t – Algebraic sum of horizontally resolved forces (thrust force)

Because F_c and F_t are acting in perpendicular plane, these forces can be measured by using dynamometer.

Merchant circle is used for determining the actual forces from measurable forces.

Merchant suggested a compact and easiest way of representing the various forces inside a circle having the vector R as diameter.

Fig. 2.17 shows merchant circle diagram which is convenient to determine the relation between the various forces and angles.

From Merchant circle of forces

$$\sin \beta = \dfrac{F}{R} \quad \cos \beta = \dfrac{N}{R}$$

$$\sin (\beta - \alpha) = \dfrac{F_t}{R} \quad \cos (\beta - \alpha) = \dfrac{F_c}{R}$$

$$\sin (\phi + \beta - \alpha) = \dfrac{F_n}{R}, \quad \cos (\phi + \beta - \alpha) = \dfrac{F_s}{R}$$

Therefore

$$R = \dfrac{F_c}{\cos (\beta - \alpha)} = \dfrac{F_t}{\sin (\phi + \beta)}$$

$$= \dfrac{F}{\sin \beta} = \dfrac{N}{\cos \beta}$$

$$= \dfrac{F_n}{\sin (\phi + \beta - \alpha)}$$

$$= \dfrac{F_s}{\cos (\phi + \beta - \alpha)}$$

Assumptions

- Tool edge is sharp.
- The work material undergoes deformation across a thin shear plane.
- There is uniform distribution of normal and shear stress on the shear plane.
- The work material is rigid and perfectly plastic.
- The shear angle ϕ adjusts itself to give minimum work.
- The friction angle β remains constant and is independent of ϕ.
- The chip width remains constant.

2.12 RELATIONSHIP BETWEEN VARIOUS FORCES

1. Frictional Force System

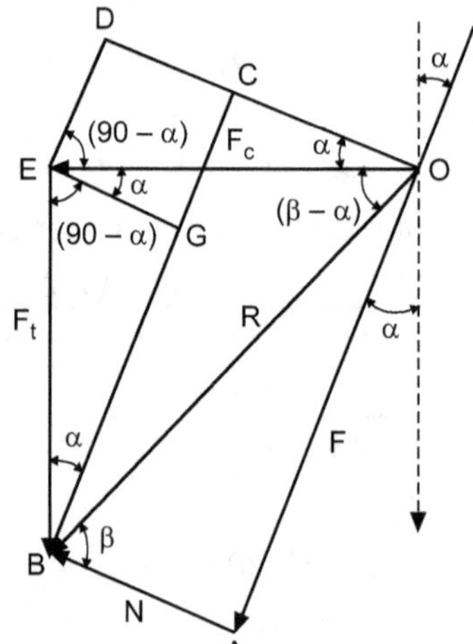

Fig. 2.18 : Forces acting in orthogonal cutting

$$F = OA = CB = CG + GB = ED + GB$$

$$F = F_c \sin \alpha + F_t \cos \alpha$$

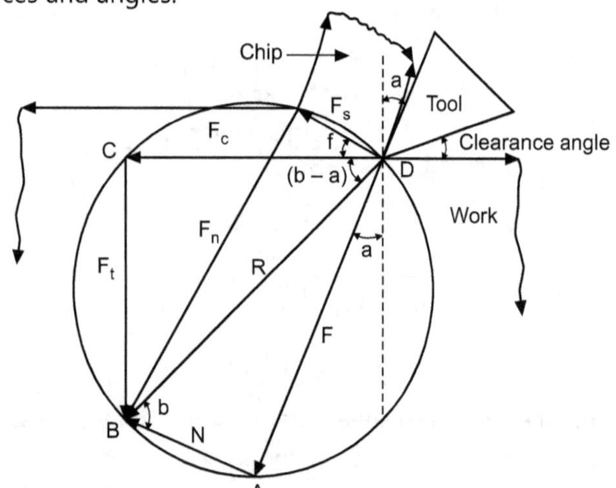

Fig. 2.17 : Merchant's circle of forces

$$N = AB = OD - CD = OD - GE$$
$$N = F_C \cos \alpha - F_t \sin \alpha$$

where, μ – coefficient of friction between chip tool interfaces.

$$\mu = \tan \beta$$
$$\tan (\beta - \alpha) = \frac{F_t}{F_c},$$
$$\tan \beta = \frac{(F_t + F_c \tan \alpha)}{(F_c - F_t \tan \alpha)}$$

The coefficient of friction,

$$\mu = \tan \beta = \frac{F}{N}$$

where, β = friction angle

In general $F_c > F_t$

but in some cases $F_t > F_c$ – broaching, grinding,

$$\frac{F_t}{F_c} = 2.5 \text{ in grinding operation.}$$

Cutting Power

The cutting power is the product of cutting speed (V_c) and cutting force (F_c).

$\therefore$ Cutting power $= V_c \times F_c$ (kw)

Specific cutting energy $= \dfrac{F_c}{b \times t}$

2. Shear Force System

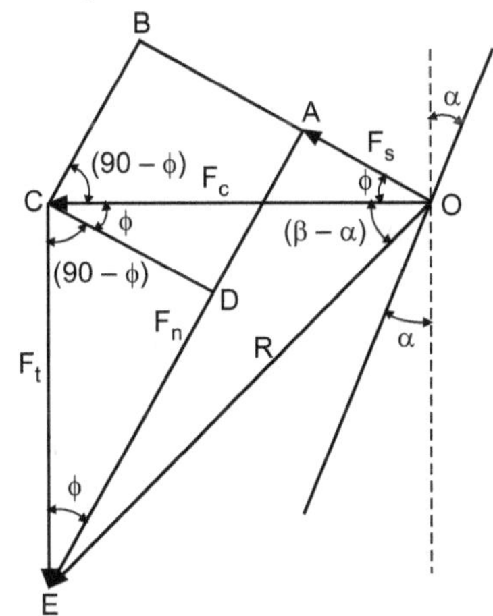

Fig. 2.19 : Forces acting in orthogonal cutting

$$F_S = OA = OB - AB = OB - CD$$
$$F_S = F_c \cos \phi - F_t \sin \phi$$
$$F_N = AE = AD + DE = BC + DE$$
$$F_N = F_c \sin \phi + F_t \cos \phi$$

also, $F_N = F_s \tan (\phi + \beta \, \alpha)$

Now, Major forces as follows

$$F = F_c \sin \alpha + F_t \cos \alpha$$
$$N = F_c \cos \alpha - F_t \sin \alpha$$

$$F_S = F_C \cos \phi - F_t \sin \phi$$
$$F_N = F_C \sin \phi + F_t \cos \phi$$
$$F_N = F_S \tan (\phi + \beta - \alpha)$$

Stress in the Chip

A chip is supposed to experience both the stress and strain during machining because it is produced as result of plastic deformation of the method.

$$F_s = \frac{[F_c \cos (\phi + \beta - \alpha)]}{[\cos (\beta - \alpha)]}$$

Shear stress $\sigma_s = \dfrac{\text{shear force}}{\text{shear area}}$

$$= \frac{F_s}{A_s} = \frac{(F_s \times \sin \phi)}{(b \times t_1)} \qquad \dots \text{N/mm}^2$$

Stress normal to shear plane (σ_n)

$$\sigma_n = \frac{F_n}{A_s}$$
$$= \frac{(F_n \times \sin \phi)}{(b \times t_1)} \qquad \dots \text{N/mm}^2$$

2.13   SPECIFIC CUTTING ENERGY (OR UNIT POWER)

Energy required to remove a unit volume of material (often quoted as a function of workpiece material, tool and process:

$$U_t = \frac{\text{Energy}}{\text{Volume removed}}$$
$$= \frac{\text{Energy per unit time}}{\text{Volume removed per unit time}}$$
$$U_t = \frac{\text{Cutting power } (P_c)}{\text{Material removal rate (MRR)}}$$
$$= \frac{F_c V}{V w t_o} = \frac{F_c}{w t_o}$$

Specific energy for shearing

$$U_s = \frac{F_s V_s}{V w t_o}$$

Specific energy for friction

$$U_f = \frac{F V_c}{V w t_o} = \frac{Fr}{w t_o}$$

Where F_C = Main Cutting force

V = Cutting velocity

W = Width of chip

t_0 = uncut chip thickness

F_S = Shear force

V_S = Shear velocity

F_r = Frictional force

2.14 PLOWING FORCE AND SIZE EFFECT

The resultant too force in metal cutting is distributed over the areas of the tool that contact the chip and workpiece. No cutting tool is perfectly sharp, and in the idealized picture as shown in Fig. 2.20. The cutting edge is represented by a cylindrical surface joining the tool flank and tool face.

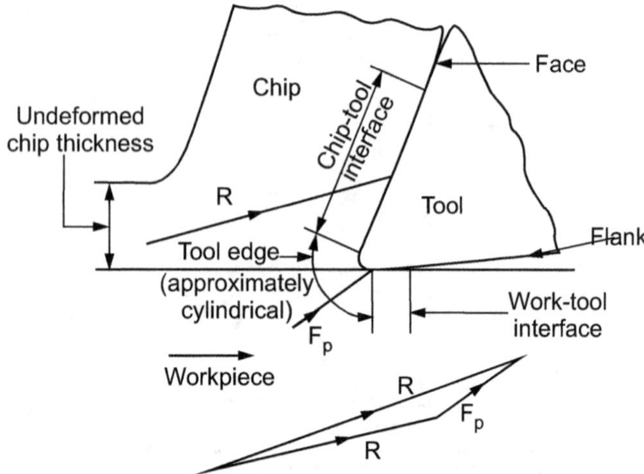

Fig. 2.20 : Plowing force and contact regions on a cutting tool

- Observation in past have shown that the radius of this edge varies from 0.005 to 0.03mm for a freshly ground high speed steel tool (HSS). As the tool edge "plows" its way through the work material, the force that acts on the tool cutting edge forms only a small proportion of the cutting force at large values of the undeformed chip thickness h_c. At the small values of h_c, however, the force that acts on the tool edge is proportionately large and cannot be neglected.

- Because of high stresses acting near the cutting edge, deformation of the tool material may occur in this region. This deformation would cause contact between the tool and new workpiece surface over a small area of the tool flank. Thus, when sharp cutting tools are used, a frictional force may act in the tool-flank region, this force again forming a small proportion of the cutting force at high feeds.

- Neither the force acting on the tool edge nor the force that may act on the tool flank contributes to removal of chip and these forces will be referred to collectively as plowing force. F_p

- The existence of the plowing force results in certain important effects and can explain the so called **size effect.** This term refers to the increase in specific cutting energy (energy required to remove a unit volume of metal) at low values of undeformed chip thickness. For example in Fig. 2.21. The mean specific cutting energy is plotted against undeformed chip

thickness for a slab milling operation. At the relatively small values of chip thickness occurring in this process, the specific cutting energy e_c increases rapidly with decreasing chip thickness.

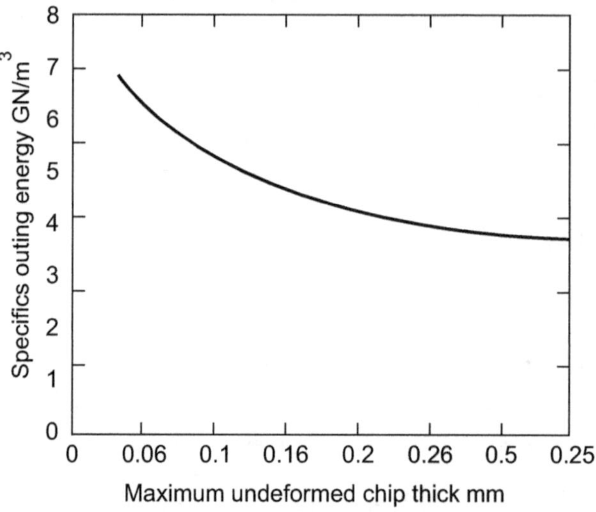

Fig. 2.21 : Effect of maximum undeformed chip thickness on specific cutting energy during slab milling, where the material is steel

- It is thought that the plowing force is constant and therefore becomes a greater proportion of the total cutting force as the chip thickness decreases. When the total cutting force is divided by undeformed chip thickness cross sectional are to give e_c, the portion of ec contributing to chip removal will remain constant , and portion resulting from plowing force F_p will increases as the chip thickness decreases explains why those processes such as grinding (as shown in Fig. 2.22) that produces very thin chips require greater power to remove a given volume of metal.

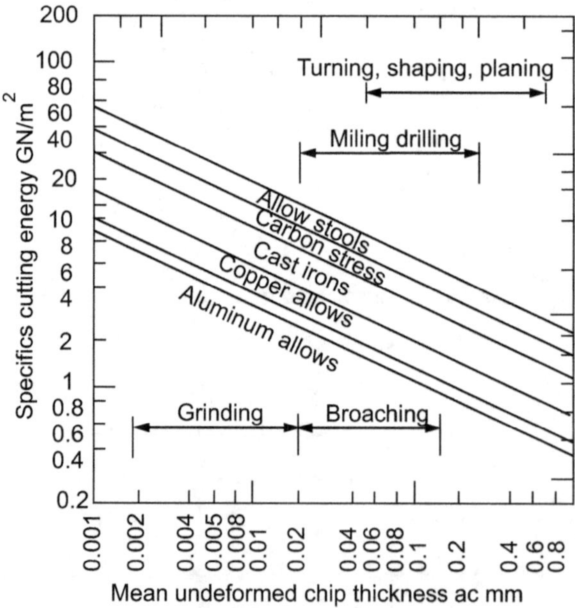

Fig. 2.22 : Approximate values of the specific cutting energy for various material and operations

2.15 MEAN SHEAR STRENGTH OF THE WORK MATERIAL

Let S be the shear strength of work material.

Shear Stress

- Shear stress acting along the shear plane

$$S = \frac{F_s}{A_s}$$

where A_s = area of the shear plane

$$A_s = \frac{t_1 w}{\sin \theta}$$

Shear stress = shear strength of work material during cutting

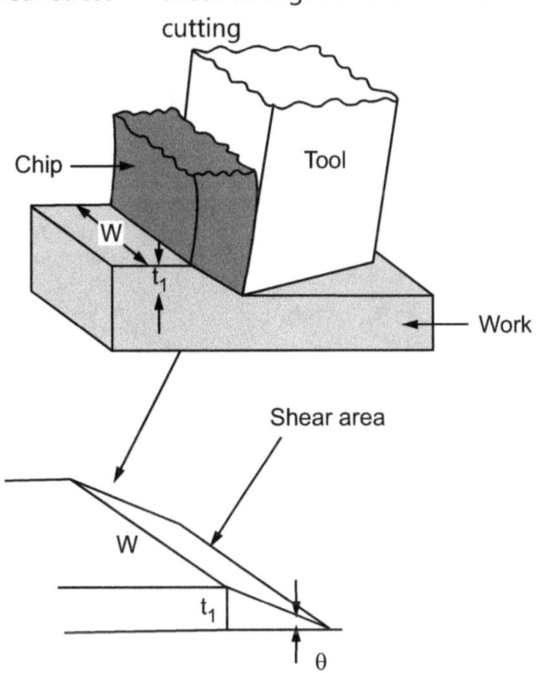

Fig. 2.23

2.16 THEORY OF ERNST AND MERCHANT (1944)

Ernst and Merchant gave the relation,

$$\phi = \frac{\pi}{4} - \frac{\beta}{2} + \frac{\alpha}{2}$$

Derivation

We know that, $F_s = \sigma_s \times A_s$

$$F_s = \frac{bt_1 \sigma_s}{\sin\phi} \qquad \ldots (2.3)$$

From merchant force circle,

$$F_c = \frac{F_s \cos(\beta - \alpha)}{\cos(\phi + \beta - \alpha)} \qquad \ldots (2.4)$$

Substitute equation (2.3) in equation (2.4), we get

$$F_c = \frac{bt_1 \sigma_s \cos(\beta - \alpha)}{\sin\phi} \left[\frac{1}{\cos(\phi + \beta - \alpha)\,\alpha} \right]$$

and power consumption,

$$P = F_c \times V_c$$

$$= V_c \times \frac{bt_1 F_s \cos(\beta - \alpha)}{\sin\phi \cos(\phi + \beta - \alpha)} \qquad \ldots (2.5)$$

During a cutting operation, ϕ takes a value such that the least amount of energy is consumed or P is minimum.

Here, V_c, b, t_1 and α are given (constant) and if we assume that F_s and β do not change when ϕ varies and that P is a function of ϕ and is of the form,

$$P(\phi) = \frac{\text{constant}}{\sin\phi \cos(\phi + \beta - \alpha)} \qquad \ldots (2.6)$$

$P(\phi)$ will be minimum when the denominator is maximum, then differentiating the denominator w.r.t ϕ and equating it to zero, we get,

$$\frac{d(\sin\phi \cos(\phi + \beta - \alpha))}{d\phi} = 0$$

$$\cos\phi \cos(\phi + \beta - \alpha) - \sin\phi \sin(\phi + \beta - \alpha) = 0$$

$$\cos(2A - B) = \cos(A + A - B)$$

$$= \cos A \cos(A - B) - \sin A \sin(A - B)$$

$$\cos(2\phi + \beta - \alpha) = 0$$

$$2\phi + \beta - \alpha = \frac{\pi}{2}$$

$$\phi = \frac{\pi}{4} - \frac{\beta}{2} + \frac{\alpha}{2} \qquad \ldots \text{hence proved}$$

2.17 THEORY OF LEE AND SHAFFER

The theory of Lee and Shaffer is based on slip line field theory and applies simplified plasticity analysis to the problem of orthogonal metal cutting. The following assumptions are made on the behaviour of work material under stress.

1. The material is rigid plastic. The stress-strain curve is shown in Fig. 2.24.

2. The behaviour of material is independent of rate of deformation.

3. The effect of temperature increase during deformation are negligible.

4. The inertia effects resulting from acceleration of material during deformation are negligible.

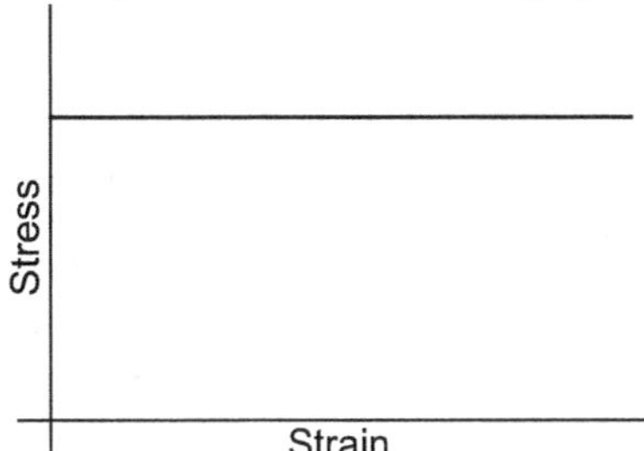

Fig. 2.24 : Stress-stain curve for a rigid-plastic material

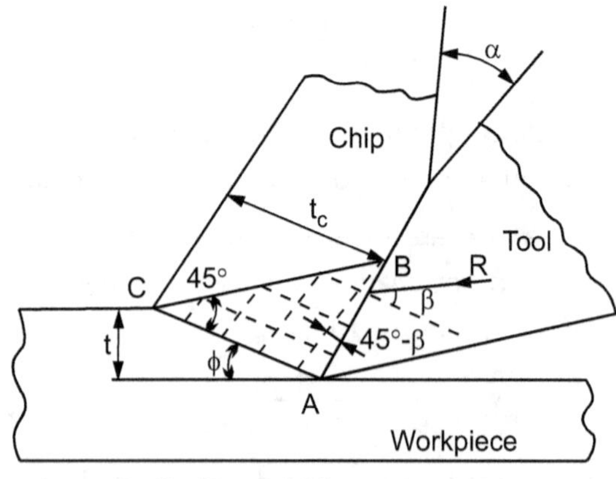

Fig. 2.25 Slip-line field for orthogonal cutting

The slip-line field for orthogonal cutting is shown in Fig. 2.25.

R = resultant tool force

$β$ = mean friction angle on tool face

$φ$ = shear plane angle

$α$ = rake angle

t = undeformed chip thickness

t_c = chip thickness

Triangle ABC contains the zone of deformation. The plane CB is stress free and slip lines meet CB at 45°. AC is the shear plane and there is velocity discontinuity along AC. A set of slip lines are parallel to AC and other set perpendicular to AC and inclined at an angle (45° − β) with the tool rake face.

From Fig. 2.25,

$$3\ CAB = 45° + β$$
$$φ + 3\ CAB = 90° + α$$
$$φ = 45° - (β - α).$$

or, $φ + β - α = 45°$ ...(2.7)

Equation (2.7) is the required shear-angle solution.

Case 1: Zero rake angle

If $β = 45°$ and $α = 0°$, $φ = 0°$

High friction and low rake angle leads to formation of built-up edge.

Case 2: Negative rake angle

If $β = 35°$ and $α = -10°$, $φ = 0°$. This solution is again impracticable.

SOLVED EXAMPLES

Example 2.1 : *A tool with 18° rake angle is making an orthogonal cut, 3 mm wide, at a speed of 31 rpm and feed of 0.25 mm. The chip thickness ratio is 0.55, cutting force is 1392 N and feed force as 363 N. Find (i) Chip thickness (ii) Shear plane angle, (iii) Coefficient of friction on tool face (iv) Shear force, (v) Energy consumed in kw min per cubic centimeter of metal removed.*

Solution : Given, $α = 18°$, $t = b = 3$ mm, $V_c = 31$ rpm, $f = 0.25$ mm $= t_1$, $r = 0.55$, $f_c = 1392$ N, $f_r = 363$ N

(i)   Chip thickness,

$$\frac{t_1}{t_2} = r \Rightarrow \frac{0.25}{t_2} = 0.55 \Rightarrow t_2 = 0.45 \text{ mm}$$

(ii) Shear plane angle ($φ$),

$$\tan φ = \frac{r \cos α}{1 - r \sin α} = \frac{0.55 \cos (18)}{1 - 0.55 \sin (18)}$$

∴ $φ = 32.22$

(iii) Coefficient of friction,

$$μ = \frac{f_t + f_c \tan α}{f_c - f_t \tan α} = \frac{363 + 1392 \tan (18)}{1392 - 363 \tan (18)}$$

∴ $μ = 0.64$

(iv) Shear force (f_s),

$$f_s = f_c \cos φ - f_t \sin φ$$
$$= 1392 \cos (32.22) - 363 \sin (32.22)$$
$$f_s = 984.1 \text{ N}$$

(v) Energy consumed $= \dfrac{\text{Power}}{\text{MRR}}$

$$= \frac{f_c \times V_c}{f \times t \times V_c} \times \frac{1}{1000}$$
$$= \frac{1392 \times 31}{1000 \times 60} \times \frac{1}{0.025 \times 0.3 \times 3100}$$
$$E = 0.0309 \left(\frac{\text{kw min}}{\text{cm}^3}\right)$$

Example 2.2 : *In an orthogonal cutting with a tool rank angle 10°, the following observations were made.*

Chip thickness ratio = 0.4, $f_t = 1200$ N, $f_c = 1600$ N, from merchants theory, calculate.

(i) Shear plane angle (ii) Shear force (iii) Normal force (iv) Coefficient of friction (v) Friction angle (vi) Resultant cutting force.

Solution : Given, $α = 10°$, $r = 0.4$, $f_t = 1200$ N, $f_c = 1600$ N

(i)   Shear angle ($φ$)

$$\tan φ = \frac{r \cos α}{1 - r \sin α} = \frac{0.4 \cos 10}{1 - 0.4 \sin 10} = 22.94°$$

(ii) Shear force,

$$f_s = f_c \cos φ - f_t \sin φ$$
$$= 1600 \cos (22.94) - 1200 \sin (22.94)$$
$$= 1005.74 \text{ N}$$

(iii) Normal force,

$$f_n = f_t \cos φ + f_c \sin φ$$
$$= 1200 \cos (22.94) + 1600 \sin (22.94)$$
$$= 1728.72 \text{ N}$$

(iv) Coefficient of friction (μ),

$$\mu = \frac{f_c \sin \alpha + f_t \cos \alpha}{f_c \cos \alpha - f_t \sin \alpha}$$

$$\mu = \frac{1600 \sin (10) + 1200 \cos (10)}{1600 \cos (10) - 1200 \sin (10)} = 1.067$$

(v) Friction angle (β),

$$\mu \tan \beta \Rightarrow \beta = \tan^{-1} (\mu)$$

$$\therefore \quad \beta = \tan^{-1} (1.067) = 46.87°$$

(vi) Resultant force (R),

$$R = \sqrt{f_t^2 + f_c^2}$$
$$= \sqrt{1200^2 + 1600^2} = 2000 \text{ N}$$

Example 2.3 : *A 250 mm diameter bar is turned at 40 rev/min with depth of cut 2mm and feed of 0.3 mm/rev. Cutting force and feed force for this operations are 1850 N and 450 N respectively. Calculate (i) Power consumption, (ii) Specific cutting energy take MRR during turning operation is 2.5×10^4 mm^3/min.*

Solution : Given, D = 250 mm, N = 40 rev/min, feed = 0.3 mm/rev, DOC = 2 mm, F_c = 1850 N, f_T = 450 N

Cutting velocity,

$$V = \frac{\pi D N}{60 \times 1000} \text{ m/s} \qquad \text{as N = rev/min}$$

$$V = \frac{\pi \times 250 \times 45}{60 \times 1000}$$

$$V = 0.589 \text{ m/s}$$

(i) Cutting power P_c

$$= \frac{f_c \times V}{1000} - \text{kw}$$

$$= \frac{1850 \times 0.589}{1000}$$

$$P_c = 1.09 \text{ kw}$$

(ii) Specific cutting energy

$$= \frac{P_c}{MRR} = \frac{1.09 \times 10^3 \times 60}{2.5 \times 10^4}$$

$$= 2.616 \text{ w.s/mm}^3$$

Example 2.4 : *A seamless tube of 50 mm outside dia. is turned on a lathe with a cutting speed of 20 m/min. The tool rake angle is 15° and feed rate is 0.2 mm/rev. The length of continuous chip in one revolution measures 80 mm. Calculate (i) Chip thickness ratio (ii) Shear plane angle (iii) shear flow velocity (iv) Shear strain (v) Shear strain rate*

Solution : Given, D = 50 mm, V_c = 20 m/min, α = 15°, f = t_1 = 0.2 mm/rev, L_2 (cut) = 80 mm/rev, L_1 – length of uncut chip $L_1 = \pi D = \pi \times 50 = 157$ m

(i) Chip thickness ratio (r) = $\dfrac{t_1}{t_2} = \dfrac{L_2}{L_1}$

$$\therefore \quad r = \frac{80}{157.07} = 0.509$$

(ii) Shear angle (ϕ), $\tan \phi = \dfrac{r \cos \alpha}{1 - r \sin \alpha} = \dfrac{0.059 \times \cos 15}{1 - 0.509 \sin 15}$

$$\therefore \quad \phi = 29.52°$$

(iii) Shear/chip flow velocity (v_f) = $V_c \times r$ = 20×0.509
$$= 10.18 \text{ m/min}$$

(iv) Shear strain (s)

$$= \frac{\cos \alpha}{\cos (\phi - \alpha) \sin (\phi)}$$

$$= \cot \phi + \tan (\phi - \alpha)$$

$$= \cot (29.52) + \tan (29.52 - 15)$$

$$= 2.025$$

(v) Shear strain rate ($\dot{S}$)

$$\frac{s}{f} = \frac{2.025}{0.2} = 10.125 \text{ rev/mm}$$

Example 2.5 : *In an orthogonal turning of steel bar with 80mm diameter on lathe with following data.* **[Feb. 15, 6M]**
Cutting speed = 110m/min, Back rate = 10°, Feed rate = 0.20 mm/rev, Cutting force = 1200 N, Feed force = 450 N, Chip thickness = 0.4 mm, Calculate : (i) Shear angle, (ii) Coefficient of friction, (iii) Cutting power.

Solution : Given,

$$V = 11 \text{ Dm/min, d = 80 mm,}$$

$$r = \frac{t_1}{t_2} = \frac{0.20}{0.4} = 0.5\text{m}$$

(i) Shear angle,

$$\tan \phi = \frac{r \cos \alpha}{1 - r \sin \alpha} = \frac{0.5 \cos(10°)}{1 - 0.5 \sin (10°)}$$

$$\phi = 28.33°$$

(ii) Coefficient of friction,

$$\mu = \frac{F_c \sin \alpha + F_c \cos \alpha}{F_c \cos \alpha - F_t \sin \alpha}$$

$$\mu = \frac{1200 \sin (10°) + 450 \cos (10°)}{1200 \cos (10°) - 450 \sin (10°)} = 0.59$$

(iii) Cutting power

$$= \frac{F_c \cdot V}{1000}, \quad \therefore V = \frac{110}{60} = 1.84 \text{ m/s}$$

$$= \frac{1200 \times 1.84}{100} = 2.21 \text{ kW}$$

Example 2.6 : *In orthogonal cutting of a 60 mm diameter MS bar on lathe, the following data was obtained:*
Rate angle = 10°, Cutting Speed = 100 m/min, Cutting force = 200N, Feed Force = 70N, Chip thickness = 0.3 mm, Feed = 0.2 mm/rev.
Calculate:
(i)   Shear angle
(ii) Coefficient of friction
(iii) Chip flow Velocity
(iv) Friction Angle   **[Nov. 16, 17; 4M]**

Solution : $\alpha = 10°$, $V = 100$ m/min, $F_c = 200$ N, $t_1 = f = 0.2$ mm/rev, $F_t = 70N$, $D = 60$ mm

$$\therefore \quad r = \frac{t_1}{t_2} = \frac{0.2}{0.3} = 0.67$$

(i) Shear angle (ϕ),

$$\tan \phi = \frac{r \cos \alpha}{1 - r \sin \alpha} = \frac{0.67 \cos 10°}{1 - 0.67 \sin 10°}$$

$$\phi = 36.75°$$

(ii) Coefficient of friction (μ)

$$\mu = \frac{F_c \sin \alpha + F_t \cos \alpha}{F_c \cos \alpha - F_t \sin \alpha}$$

$$= \frac{200 \sin 10 + 70 \cos 10}{200 \cos 10 - 70 \sin 10}$$

$$\mu = 0.56$$

(iii) Chip velocity flow (V_F)

$$\therefore \quad V_F = V \times r$$

$$= 1.67 \times 0.67 \qquad ...v = \frac{100}{60} = 1.67 \text{ m/s}$$

$$= 1.119 \text{ m/sec}$$

(iv) Friction angle (β)

$$\mu = \tan \beta$$

$$\therefore \quad 0.56 = \tan \beta$$

$$\therefore \quad \beta = 29.25°$$

Example 2.7 : *In an orthogonal cutting with a tool of rake angle 10°, the following observations were made.*

Chip thickness ratio = 0.4.

Horizontal component of the cutting force = 1200 N.

Vertical component of the cutting force = 1600 N.

From Merchant's theory, calculate :

(i) Shear plane angle.

(ii) Friction force along rake face.

(iii) Normal force on the rake face.

(iv) Coefficient of friction (μ) at the chip tool interface.

(v) Friction angle (β).

(vi) Resultant cutting force (R).

Solution : Given :

   Rake angle = 10°.

   Chip thickness ratio = 0.4.

   Horizontal cutting force F_c = 1200 N.

   Vertical component of the cutting = 1600 N.

(i) Shear plane angle ϕ

$$\phi = \tan^{-1} \left[\frac{r \cos \alpha}{1 - r \sin \alpha} \right]$$

$$= \tan^{-1} \left[\frac{0.4 \cos 10°}{1 - 0.4 \sin 10°} \right]$$

$$= \tan^{-1} \left[\frac{0.393}{0.930} \right]$$

$$= 22.90°$$

(ii) Friction angle β,

$$\mu = \tan \beta = \frac{F_c \sin \alpha + F_t \cos \alpha}{F_c \cos \alpha - F_t \sin \alpha}$$

$$= \frac{1200 \sin 10° + 1600 \cos 10°}{1200 \cos 10° - 1600 \sin 10°}$$

$$= 17.81°$$

Coefficient of friction (μ) at the chip tool interface,

$$\mu = \tan \beta = \tan 17.81°$$

$$\mu = 0.321$$

(iv) Friction force along rake face,

$$F = F_c \sin \alpha + F_t \cos \alpha$$

$$= 1200 \sin 10 + 1600 \cos 10$$

$$F = 1784.07 \text{ N}$$

(v) Normal force on rake face,

$$N = \text{Normal force}$$

$$= F_c \cos \alpha - F_t \sin \alpha$$

$$= 1200 \cos 10 - 1600 \sin 10$$

$$= 903.93 \text{ N}$$

(vi) Resultant cutting force,

$$R = F + N$$

$$R = 2688 \text{ N}$$

Example 2.8 : *In an orthogonal cutting operation, following observations were made :*

   (1) Cutting speed = 25 m/min.

   (2) Width of cut = 2.5 mm

   (3) Feed = 0.24 mm/rev.

   (4) Chip thickness = 0.4 mm

   (5) Cutting force = 1400 N

   (6) Thrust force = 400 N

   (7) Tool rake angle = 5°.

Calculate : (i) Shear angle, (ii) Friction angle, (iii) Chip velocity, (iv) Shear strain, (v) Power consumed at tool in kW-min/cm³ of metal removed.

Solution : $v = 25$ m/min $= 0.4167$ m/s

$$b = 2.5 \text{ mm}$$

$$f = 0.24 \text{ mm/rev}$$

$$t_2 = 0.4 \text{ mm}$$

$$F_c = 1400 \text{ N}$$

$$F_t = 400 \text{ N}$$

$$\alpha = 5°$$

(i) Shear angle ϕ

   Also   feed = Uncut chip thickness

$$f = t_1 = 0.24 \text{ mm}$$

$$r = \frac{t_1}{t_2} = \frac{0.24}{0.4} = 0.6$$

$$\phi = \tan^{-1}\left(\frac{r \cos \alpha}{1 - r \sin \alpha}\right)$$

$$\phi = \tan^{-1}\left(\frac{0.6 \times \cos (5°)}{1 - 0.6 \times \sin (5°)}\right)$$

$$\phi = 32.24°$$

(ii) Friction angle β

$$\mu = \tan \beta = \frac{F_C \sin \alpha + F_t \cos \alpha}{F_C \cos \alpha - F_t \sin \alpha}$$

$$= \frac{1400 \times \sin (5°) + 400 \times \cos (5°)}{1400 \times \cos (5°) - 400 \times \sin (5°)}$$

$$\tan \beta = \frac{520.49}{1359.81} = 0.3827$$

$$\beta = 20.95°$$

(iii) Chip velocity v_C

$$r = \frac{v_C}{v}$$

$$v_C = r \times v = 0.6 \times 0.4167 = 0.25 \text{ m/s}$$

(iv) Shear strain S

$$S = \cot \phi + \tan (\phi - \alpha)$$
$$= \cot (32.24°) + \tan (32.24° - 5°)$$
$$= 2.1$$

(v) Power consumed

$$\text{Power} = \frac{F_C \, v}{1000} = \frac{1400 \, (0.4167)}{1000} = 0.584 \text{ kW}$$

Metal removal rate $= A_C \times v$

where $A_C = b \times t = 2.5 \, (0.24) = 0.6 \text{ mm}^2$

$$A_C = 6 \times 10^{-3} \text{ cm}^2$$

$$v = 25 \times 100 = 2500 \text{ cm/min}$$

$\therefore$ Metal removal rate $= 6 \times 10^{-3} \, (2500)$

$$= 15 \text{ cm}^3/\text{min}$$

Power consumed in kW-min/cm^3 $= \dfrac{0.584}{15}$

$$= 0.03894 \text{ kW-min/cm}^3$$

Example 2.9 : *A seamless tube of 50 mm outside diameter is turned on a lathe with a cutting speed of 20 m/min. The tool rake angle is 15° and feed rate is 0.2 mm/rev. The length of continuous chip in one revolution measures 80 min. Calculate :*

(i) Chip thickness ratio

(ii) Shear plane angle

(iii) Shear flow speed

(iv) Shear strain

(v) Shear strain rate.

Solution : $d = 50$ mm

$$v = 20 \text{ m/min}$$
$$\alpha = 15°$$
$$t_1 = 0.2 \text{ mm}$$
$$lc/rev = 80 \text{ mm/rev}$$

$$v = \frac{\pi d N}{60} \times 10^{-3} \text{ m/min}$$

$\therefore$ $20 = \pi \times 50 \times N \times 10^{-3}$

$\therefore$ $N = 127.32395$ r.p.m.

$\therefore$ $v_C = 80 \text{ mm/rev} \times 127.32395 \text{ rev/min}$

$$= 10185.916 \text{ mm/min}$$
$$= 10.185916 \text{ m/min}$$

(i) Chip thickness ratio (r)

$$r = \frac{t_1}{t_2} = \frac{v_C}{v} = \frac{10.185916}{20}$$

$$r = 0.5092958$$

(ii) Shear plane angle ϕ

$$\tan \phi = \frac{r \cos \alpha}{1 - r \sin \alpha} = \frac{0.5092 \cos 15°}{1 - 0.5092 \sin 15°}$$

$\therefore$ $\tan \phi = 0.5666329$

$\therefore$ $\phi = \tan^{-1} (0.5666329)$

$\therefore$ $\phi = 29.537321°$

(iii) Shear flow speed

$$\frac{v_S}{v} = \frac{\cos \alpha}{\cos (\phi - \alpha)}$$

$\therefore$ $v_S = \dfrac{v \cos \alpha}{\cos (\phi - \alpha)} = \dfrac{20 \cos 15°}{\cos (29.537° - 15°)}$

$\therefore$ $v_S = 19.957468$ m/min

Shear strain (S)

$$S = \cot \phi + \tan (\phi - \alpha)$$
$$= \cot (29.537°) + \tan (29.537° - 15°)$$
$$S = 2.0241235$$

Shear plane length $= \dfrac{t_1}{\sin \phi} = \dfrac{0.2}{\sin (29.53°)}$

$$= 0.4056874 \text{ mm}$$

$\therefore$ $t_S = \dfrac{1}{10} \times$ shear plane length

$$= 0.04056874 \text{ mm}$$

(v) Shear strain rate $= \dfrac{v_S}{t_S}$

$$v_S = 19.957468 \text{ m/min} = 0.3326244 \text{ m/s}$$

$$= 0.332644 \times 10^3 \text{ mm/s}$$

Shear strain rate $= \dfrac{0.3326 \times 10^3}{0.04056874} \text{ s}^{-1}$

$$= 8.199033 \times 10^3 \text{ s}^{-1}$$

Example 2.10 : *In an orthogonal cutting operations, following are the observations :*

    (i) Cutting speed = 120 m/min.

    (ii) Uncut chip thickness = 0.127 mm

    (iii) Rake angle = 10°

    (iv) Width of cut = 6.35 mm

    (v) Cutting force = 567 N

    (vi) Thrust force = 227 N

    (vii) Chip thickness = 0.228 mm.

Calculate : Shear angle, friction angle, shear stress along the shear plane, chip velocity, shear strain, cutting power.

Solution :

$$v = 120 \text{ m/min} = 2 \text{ m/sec}$$
$$t_1 = 0.127 \text{ mm}$$
$$t_2 = 0.228 \text{ mm}$$
$$\alpha = 10°$$
$$b = 6.35 \text{ mm}$$
$$F_C = 567 \text{ N}$$
$$F_t = 227 \text{ N}$$

(i) $r = \text{chip thickness ratio} = \dfrac{t_1}{t_2} = \dfrac{0.127}{0.228} = 0.557.$

(ii)   Shear angle $= \phi$

$$\tan \phi = \frac{r \cos \alpha}{1 - r \sin \alpha}$$

$$\phi = \tan^{-1}\left(\frac{r \cos \alpha}{1 - r \sin \alpha}\right)$$

$$= \tan^{-1}\left(\frac{0.557 \times \cos 10}{1 - 0.557 \sin 10}\right)$$

$$\phi = 31.26°$$

(iii) Friction angle

$$= \tan \beta = \frac{F_C \sin \alpha + F_t \cos \alpha}{F_C \cos \alpha - F_t \sin \alpha}$$

$$= \frac{567 \sin 10 + 227 \times \cos 10}{567 \cos 10 - 227 \times \sin 10}$$

$$= 0.6204$$

$$\beta = \tan^{-1}(0.6204)$$

$$\beta = 31.81°$$

(iv) Shear stress along shear plane,

$$\tau_S = \frac{F_S}{A_S} = \frac{F_S \cdot \sin \phi}{b \cdot t_1}$$

$$\therefore \quad F_S = F_C \cos \phi - F_t \sin \phi$$

$$= 567 \times \cos 31.26 - 227 \sin 31.26$$

$$= 366.88 \text{ N}$$

$$\tau_S = \frac{366.88 \times \sin 31.26}{6.35 \times 0.127}$$

$$= 236.07 \text{ N/mm}^2$$

(v)   Chip velocity $= v_C = v \cdot r$

$$= 2 \times 0.557$$

$$v_C = 1.114 \text{ m/sec}$$

(vi) Shear strain $= e = \cot \phi + \tan(\phi - \alpha)$

$$= \cot 31.26 + \tan(31.26 - 10)$$

$$= 1.6473 + 0.3890$$

$$= 2.036$$

(vii) Power consumed

$$= \frac{F_C \, v}{1000}$$

$$= \frac{567 \times 2}{1000} = 1.134 \text{ kW}$$

EXERCISE

1. What type of surfaces can be produced by a milling process?

2. Explain the two basic types of milling operations.

3. Sketch and contrast the two milling methods of machining flat surfaces.

4. What is the basic type of milling machine?

5. Sketch and explain the working of a plain column and knee type milling machine.

6. Sketch and describe a vertical milling machine.

7. What is the difference between a fixed head and a swivelling head vertical milling machine?

8. Describe the difference between a horizontal milling machine and a vertical milling machine.

9. What is feed on a milling machine?

10. How is the size of a column and knee type milling machine given?

11. Sketch and describe a Bed type milling machine.

◈ ◈ ◈

THERMAL ASPECTS, TOOL WEAR, AND MACHINABILITY

3.1 HEAT GENERATION IN METAL CUTTING

The temperature generated during the metal cutting controls the rate of tool wear, the practical cutting speed and the metal removed rate.

1. Primary Deformation Zone (Chip - Work Interface)

- It is the region in which actual plastic deformation of the metal occurs during machining. Due to this deformation heat is generated.
- A portion of heat is carried away by the chip; due to which its temperature is raised.
- The rest of the heat is retained by the work piece. It is known as primary deformation zone.

2. Secondary Deformation Zone (Tool Chip Interface)

- As the chip slides upwards along the face of the tool friction occurs between their surfaces, due to which heat is generated.
- A part this heat is carried the chip, which further raises the temp of the chip.
- This area is known as secondary deformation zone.

3. Tool- WORK PIECE INTERFACE

- This portion of tool flank which rubs against the work surface is another source of heat generation due to friction.
- This heat is also shared by the tool, workpiece and the coolant used. It is more pronounced when the tool is not sufficiently sharp.
- Chip carries 70 % of heat. Work piece carries 15% of heat and tool carries the remaining 15% of heat generated.

3.2 TEMPERATURE DISTRIBUTION IN METAL CUTTING

- It is hard to predict the intensity and distribution of heat sources in individual machining operations because properties of material used in machining vary with temperature, the mechanical process and thermal dynamic process are tightly coupled together.

- In the elastic deformation, the energy required for the operation is stored in the material as strain energy and no heat is generated. However, in case of plastic deformation, most of the energy used is converted into heat.

There are three sources/regions for heat generation in metal cutting,

1. Primary deformation zone
2. Tool chip interface zone or Secondary deformation zone
3. Tool work piece interface zone.

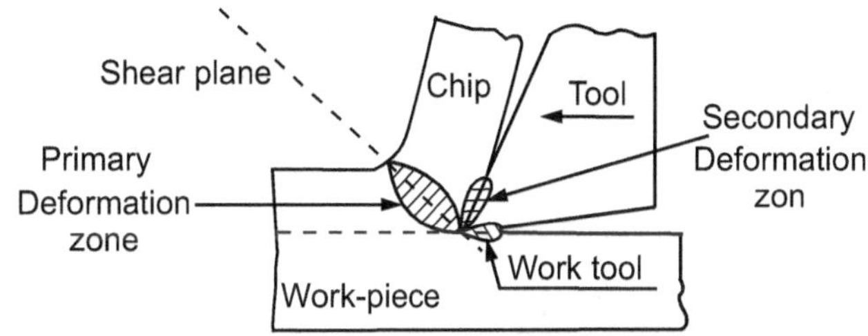

Fig. 3.1 : Source of heat generation in machining

- Heat generated in primary zone is mainly due to plastic deformation and viscous dissipation.

- In secondary zone, heat is generated due to chip distortion. Metal layer inside the bend and adjacent to tool tip is stretched while l ayer away from the tip is contracted. Hence, there is distortion of gains in the chip which leads to internal friction between heated chip and tool face.

- In Work tool interface zone, heat is generated due to friction by rubbing action between portion of tool flank and work piece.

3.3 MEASUREMENT OF CUTTING FORCES

- Measurement of cutting forces in machining is useful for determination of power requirements on machine tools, proper design of tool holders, jigs and fixtures and other parts.

- Important and generally accepted method of measurement of forces is by cutting tool dynamometers.

- Tool dynamometers for measurement of cutting forces may be mechanical, electrical, pneumatic or other types.

- The dynamometers must be sensitive, rigid, insensitive to humidity and temperature variation, consistent and repeatable.

The Dynamometers have Three Basic Parts

1. A detector transducer to note change of state by picking signal.

2. Signal magnification device.

3. Display system for output reading.

3.3.1 Turning Dynamometer

- Strain gauge dynamometers are most commonly used. The basic principle is shown in Fig. 3.2.

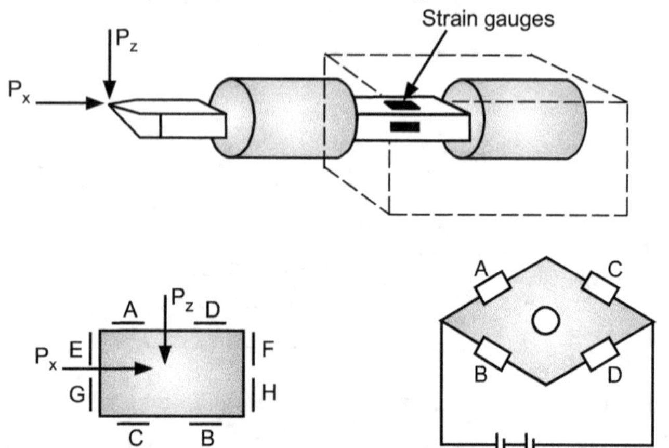

Fig. 3.2 : Dynamometer

- The two forces are measured by this scheme. The pairs of strain gauges are placed in the direction of forces to be measured.

- The strain gauge consists of a grid of fine (0.025 mm) wire bonded between the two thin insulating strips of paper.

- Each end of the grid is wired into a Wheatstone bridge circuit. The strain gauge assembly is then permanently cemented to either the cutting tool or member that gets strained on application of force.

- Upon being strained, the wire cross-section of strain gauge is changed. Its resistance to change is in direct proportion to the strain or force being measured.

- The unbalance of Wheatstone bridge indicates the cutting forces.

- Each of the two sets of four gauges are connected in two bridge circuits to enable the determination of P_x and P_z. With proper calibration, accurate measurement of forces is obtained.

- In drilling operation, measurement is done for thrust and torque. Milling involves generation of three forces.

3.3.2 Cutting Forces

- The forces acting on a single point cutting tool while performing turning operation are shown in Fig. 3.3.

➢ Feed force F_x in the direction opposite to feed.

➢ Thrust force F_y acting in direction perpendicular to surface.

➢ Cutting force (tangential) F_z in the direction of main cutting motion.

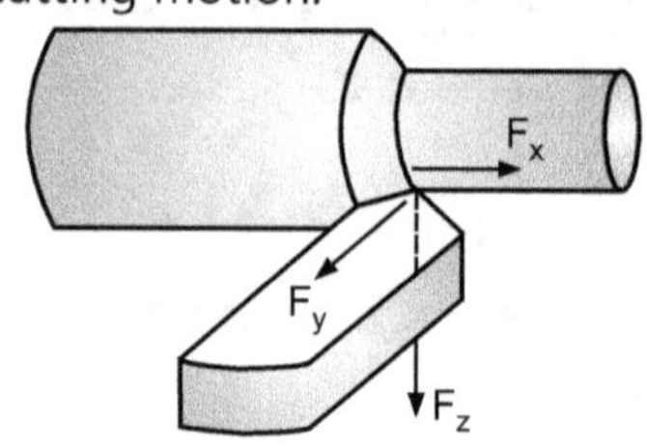

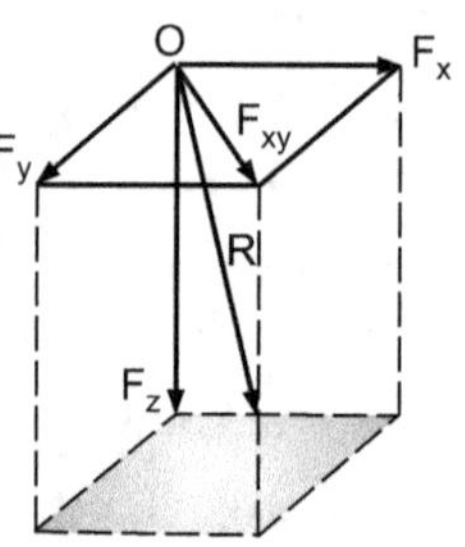

Fig. 3.3

$$R = \text{Resultant force} = \sqrt{F_x^2 + F_y^2 + F_z^2}$$

3.3.3 Cutting Speed, Feed, Depth of Cut

- **Cutting Speed :** It is travel of a point on cutting edge relative to the surface of cut in unit time. In lathe machine turning operation, cutting speed, $v = \dfrac{\pi DN}{1000}$ m/min, where D is diameter of work in mm, N = rpm of job.

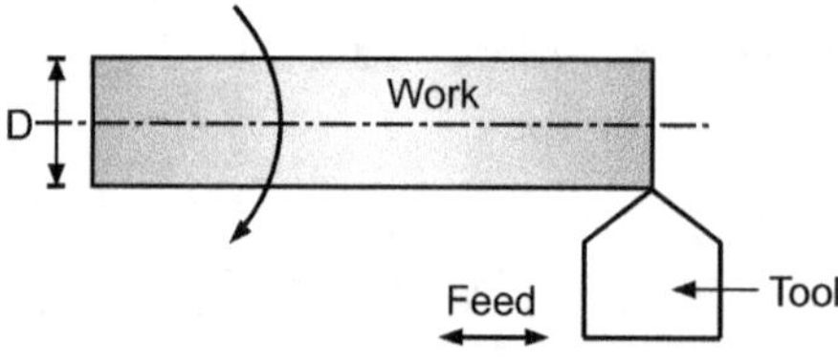

Fig. 3.4

- **Feed :** Amount of tool advance per revolution of job parallel to the surface being machined. Unit is mm/rev.

- **Depth of Cut :** It is the thickness of layer of metal removed in one cut to pass measured in direction perpendicular to the machined surface. Refer Fig. 3.5.

$$\text{Depth of cut, } t = \frac{D - d}{2} \text{ mm}$$

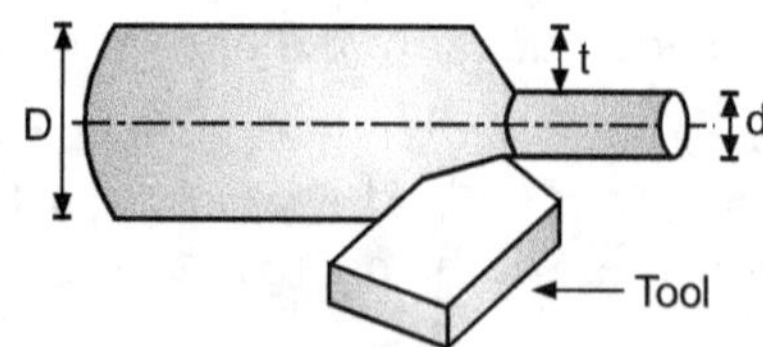

Fig. 3.5

Above Three Cutting Parameters Depend on

- Type of material of workpiece and cutting tool.
- Quality of surface required.
- Type of machine and coolant used.
- Type of operation performed.

3.3.4 Effect of Cutting Conditions on Cutting Forces

Cutting Speed

- For lower cutting speeds, the force required to cut is higher and as cutting speed increases, the force required for cutting is reduced as it is constant further. (See Fig. 3.6).

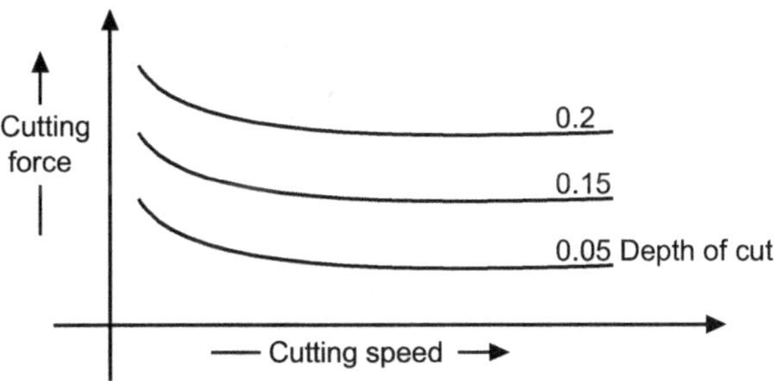

Fig. 3.6

Feed and Depth of Cut

- The cutting force does not vary directly as the cross-sectional area of cut. The specific cutting force decreases as the undeformed chip thickness increases.

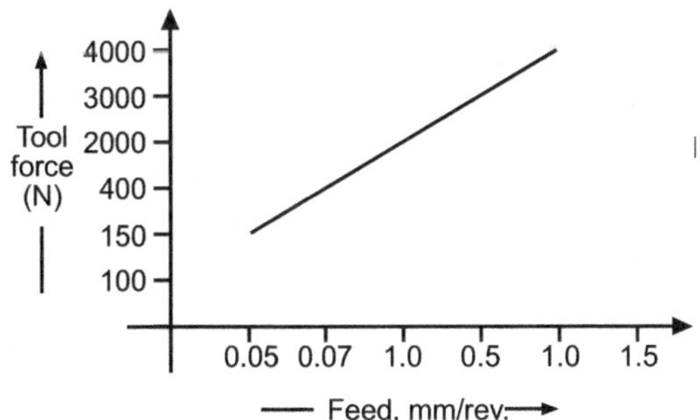

(a) Constant depth of cut

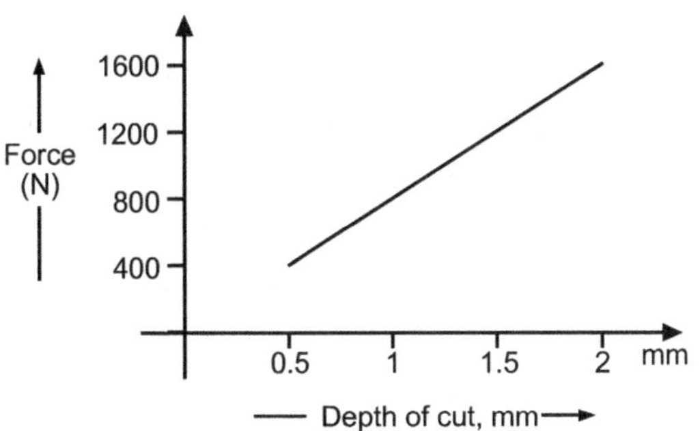

(b) Constant feed

Fig. 3.7

- Fig. 3.7 shows effect of feed and depth and cut on tool force when turning 0.21 carbon steel with HSS tool. It is difficult to calculate exact magnitude of forces acting on tool.

Empirical Formula

$$F_z = C_p f^x t^y k$$

where, C_p = coefficient, depends on tool and work

$\quad\quad f$ = feed, mm/rev.

$\quad\quad t$ = depth of cut, mm

$\quad\quad k$ = correlation coefficient = 0.9 to 1.0

$\quad x, y$ = exponents

$\quad\quad F_z$ = main or tangential force (N)

E.g. for steel having hardness BHN 215, C_p = 2250, x = 0.75 and y = 1.0 for turning and boring operations.

The components of forces F_x and F_y are,

$$F_x = (0.2 \text{ to } 0.3)\, F_z$$
$$F_y = (0.1 \text{ to } 0.2)\, F_z$$

Effect of Rake Angle and Effect of Side Cutting Edge Angle on Cutting Forces

- With increase in rake angle, the cutting force reduces. Cutting force is inversely proportional to rake angle. (See Fig. 3.8).

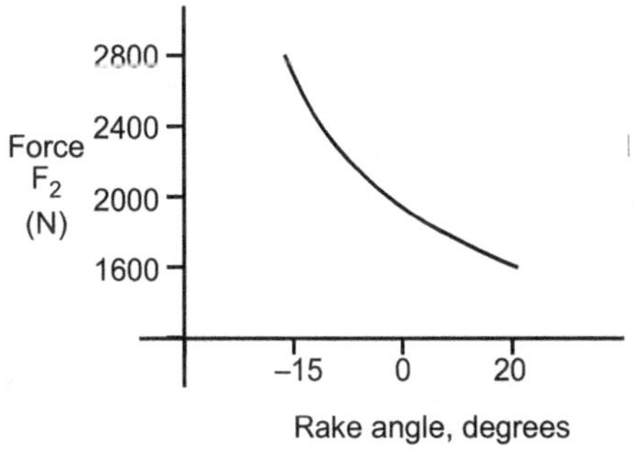

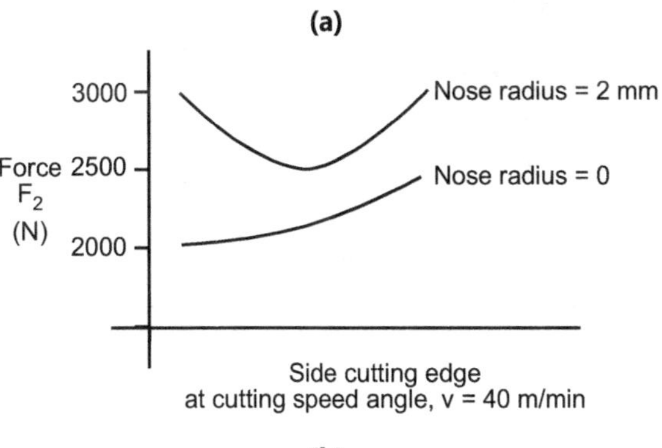

Fig. 3.8

Note : BHN – Brinell Hardness Number.

3.3.5 Effect of Various Factors on Cutting Temperature

Material of Tool and Work:

- Higher hardness and tensile strength of material required more cutting force and hence higher heat generated.

Tool Geometry:

- The rake angle has complex influence on the cutting temperature.

- If rack angle increase in positive direction both cutting force and amount of heat generated are decreased. Negative rack angle causes deformations.

- As the nose radius increase, volume of tool point is increases and it promotes more heat conduction.

- Large cross section area of tool shank leads more heat conduction and it reduces cutting temperature.

Cutting Parameter:

- As cutting speed increases, friction will increases and it raise the temperature of cutting zone. Cutting temperature is also increase with increase in feed and depth of cut.

Cutting Fluid:

- Cutting fluid is not only reduces friction but also absorb and carry away heat from work, chips and tool.

3.4 MEASUREMENT OF TEMPERATURE IN MACHINING

- There are number of methods for measuring the chip tool interface temperature as tool - work thermocouple, radiation pyrometers, embedded thermocouples, temperature sensitive paints etc.

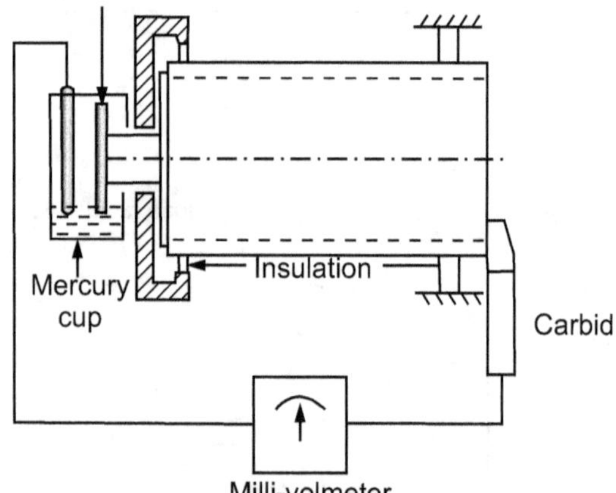

Fig. 3.9 : Tool work thermocouple

- Out of this tool work thermocouple technique is widely used because of various disadvantages of other techniques.

- Tool and work material as two elements of thermocouple. The hot junction is contact area at the cutting edge while cold part of tool forms cold junction.

- The thermoelectric *emf generated* between the tool and work piece during metal cutting is measured using sensitive milli voltmeter.

Moving Thermocouple Technique

- This simple method, schematically shown in Fig. enables measure the gradual variation in the temperature of the flowing chip before, during and immediately after its formation.

- A bead of standard thermocouple like chrome-alumel is brazed on the side surface of the layer to be removed from the work surface and the temperature is attained in terms of mV.

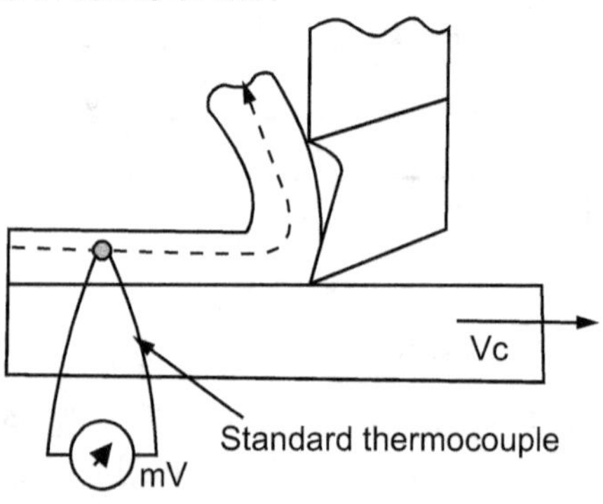

Fig. 3.10 : Moving thermocouple technique of measuring cutting temperature

3.5 NEEDS AND PURPOSES OF MEASUREMENT OF CUTTING FORCES

In machining industries and R & D sections the cutting forces are desired and required to be measured (by experiments),

- For determining the cutting forces accurately, precisely and reliably (unlike analytical method).

- For determining the magnitude of the cutting forces directly when equations are not available or adequate.

- To experimentally verify mathematical models.

- To explore and evaluate role or effects of variation of any parameters, involved in machining, on cutting forces, friction and cutting power consumption which cannot be done analytically.

- To study the machinability characteristics of any work – tool pair.

- To determine and study the shear or fracture strength of the work material under the various machining conditions.

- To directly assess the relative performance of any new work material, tool geometry, cutting fluid application and special technique in respect of cutting forces and power consumption.

- To predict the cutting tool condition (wear, chipping, fracturing, plastic deformation etc.) from the on-line measured cutting forces.

There are methods of measurement of cutting forces.

1. Indirectly

- From cutting power consumption
- By calorimetric method

Characteristics

- Inaccurate
- Average only
- Limited application possibility

2. Directly : Using tool force dynamometer(s)

Characteristics

- Accurate
- Precise / detail
- Versatile
- More reliable
- Design requirements for tool – force dynamometers.

For consistently accurate and reliable measurement, the following requirements are considered during design and construction of any tool force dynamometers :

- **Sensitivity :** The dynamometer should be reasonably sensitive for precision measurement.

- **Rigidity :** The dynamometer need to be quite rigid to withstand the forces without causing much deflection which may affect the machining condition.

- **Cross Sensitivity :** The dynamometer should be free from cross sensitivity such that one force (say P_Z) does not affect measurement of the other forces (say P_X and P_Y).

- Stability against humidity and temperature.

- Quick time response.

- High frequency response such that the readings are not affected by vibration within a reasonably high range of frequency.

- Consistency, i.e. the dynamometer should work desirably over a long period.

3.6 TOOL LIFE [Feb. May 16]

- Tool life is defined as the total cutting time accumulated before tool failure occurs.

- In general, the tool life can be defined as tool's useful life which has been expended when it can no longer produce satisfactory parts.

- The two most commonly used criteria for measuring the tool life are :

    1. Total destruction of the tool when it ceases to cut.

    2. A fixed size of wear land on tool flank.

- The tool life values as suggested by ISO are $V_B = 0.30$ mm if the flank is regularly worn in zone B. or $V_B = 0.6$ mm if flank is irregularly worn, scratched or chipped in zone B.

Tool Life can also be Measured by Following Different Criteria :

- Actual cutting time before failure.

- Length of work or number of components produced.

- Volume of metal removed before failure.

- Cutting speed for a given time to failure.

- Limiting value of surface finish.

- The advantage of wear land as failure criterion is, it is easy to measure and cutting forces and surface finish are directly proportional to it.

- The variables that affect tool life are,

    ➢ Cutting parameters - speed, feed, depth of cut.

    ➢ Tool material, work material.

    ➢ Tool geometry.

    ➢ Surface conditions of work.

    ➢ Cutting fluids.

3.6.1 Taylor's Tool Life Equation

- In 1907, Taylor gave the relationship between cutting speed and tool life,

$$VT^n = C \qquad \qquad ... (3.7)$$

where, V = cutting speed (m/min)

T = time (min) for the flank wear to reach a certain dimensions

C = constant

n = exponent, which depends upon the cutting tool material

= 0.1 to 0.15 for HSS tools

= 0.2 to 0.9 for carbide tools

= 0.4 to 0.6 for ceramic tools

- If cutting speed-tool life curves are plotted on a log-log graph, straight line is obtained. (See Fig. 3.11)

  n is the negative slope of the curve and C is the intercept velocity at T = 1.

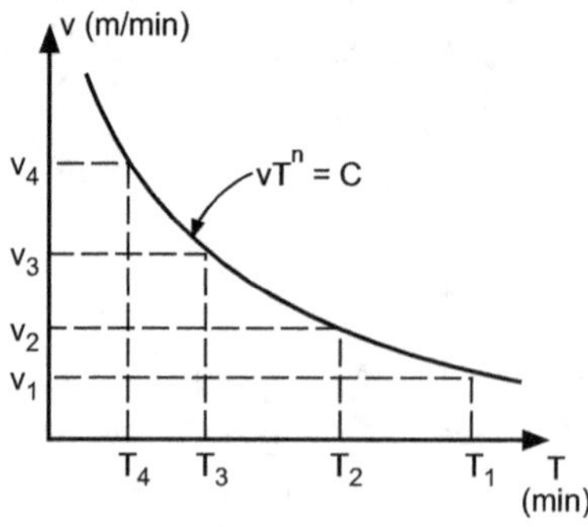

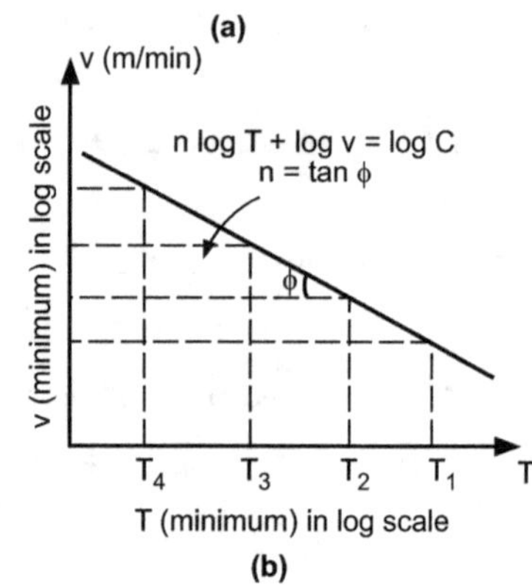

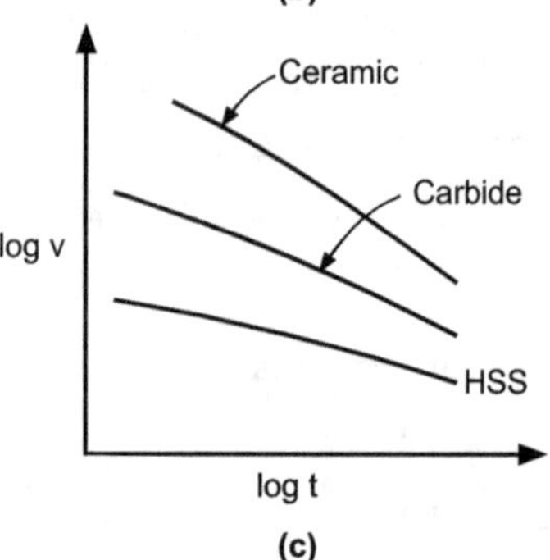

Fig. 3.11 : Relation between cutting speed (v) and tool life (T)

- The tool life also depends to a great extent on the depth of cut (d) and feed rate per revolution (f). Assuming a logarithmic variation of C with d, equation (3.7) can be written as,

$$VT^n \cdot d^m = C$$

Considering feed rate also,

$$VT^n \cdot d^m \cdot f^k = C$$

- It has been seen that decrease of life with increased speed is twice as great (exponentially) as the decrease of life with increased feed.

3.6.2   Factors on which Tool Life Depends

- **Cutting Speed** : More is the cutting speed, less is the tool life. Less is the cutting speed, more is the tool life.

- **Feed and Depth of Cut** : More is the feed and depth of cut, less is the tool life.

- **Tool Geometry** : Tool life depends upon the noise radius, rake angle, clearance angle and cutting edge angle.

- **Tool Material** : More is the hardness of tool material, more is the tool life. Like cementite, carbide is having more tool life.

- **Cutting Fluid :** If you use more cutting fluid, then friction will reduce and it will provide lubrication. Hence, tool life will increase.

3.6.3   Basic Requirements of Tool Materials

Tool material should possess following:

- **Toughness :** To avoid fracture failure

- **Hot Hardness :** Ability to retain hardness at high temperatures

- **Wear Resistance :** Hardness is the most important property to resist abrasive wear.

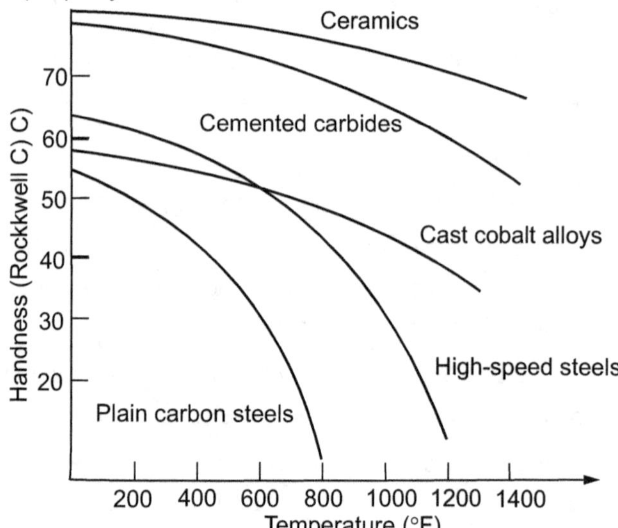

Fig. 3.12

Fig. 3.12 Typical hot hardness relationships for selected tool materials. Plain carbon steel shows a rapid loss of hardness as temperature increases. High speed steel is substantially better, while cemented carbides and ceramics are significantly harder at elevated temperatures.

3.6.4   Tool Life Criteria

Tool life can be defined by number of ways as follow.

- Complete failure of cutting edge

- Visual inspection of flank wear (or rater wear) by the machine operator

- Fingernail test across cutting edge
- Changes in sound emitted from operation
- Chips become ribbony, stringy, and difficult to dispose of
- Degradation of surface finish
- Increased power
- Workpiece count
- Cumulative cutting time

Tool Life

- Tool life represents the useful life of the tool, expressed generally in time units from the start of cut to some end point defined by a failure criterion.

Tool Life Prediction

- Taylor's tool life equation predicts tool failure based on flank wear of the tool

where

V is the cutting speed, t is the tool life,

n is Taylor exponent.

$$n = 0.125 \text{ for HSS}$$
$$n = 0.25 \text{ for Carbide}$$
$$n = 0.5 \text{ for Coated Carbide/Ceramic}$$

C is a constant given for work piece material

3.7 TOOL FAILURE (WEAR)   [Feb. 15, 17]

Whenever tool is not performing machining satisfactory it is said to be fail.

Modes of Failure (Wear)

1. Flank wear
2. Crater wear
3. Chipping off of cutting edge
4. Thermal cracking and softening.
5. Diffusion wear
6. Plastic deformation

1. Flank Wear

- This type of wear occurs in the flank below the cutting edge.
- It occurs due to abrasion btw the tool flank and the w/p excessive heat generated as a result of the same.
- The magnitude of this wear mainly depends upon the relative harnesses of the w/p tool material at the time of cutting and also the extent of strain hardness of the chip.

2. Crater Wear

- This type of wear takes place in a cutting its face, at a small distance from its cutting edge.

- This type of wear takes place while machining ductile material like steel alloys, in which continuous chip is produced.
- The resultant feature of this type of wear is the formation of a crater or a depression at the tool-chip interface.
- Higher feeds and lack of cutting fluids increases the rate of crater wear.

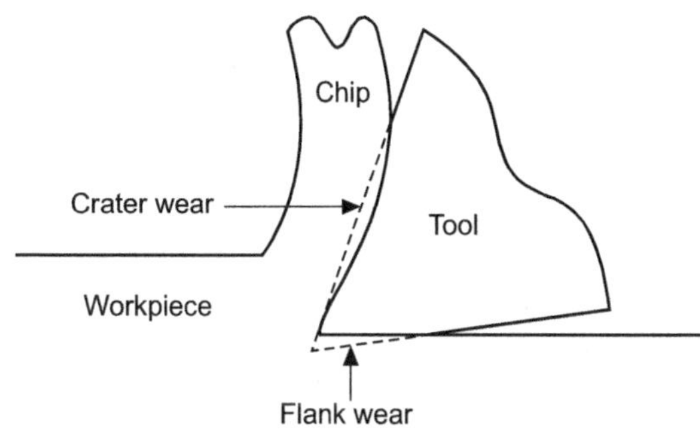

Fig. 3.13 : Tool wear

3. Chipping off of Cutting Edge

- The mechanical chipping of the nose and the cutting edge of the tool are commonly observed causes of tool failure.
- The factors responsible for this are too high cutting pressure. Mechanical impact, excessive wear, too high vibrations and chatter, weak tip and cutting edge etc

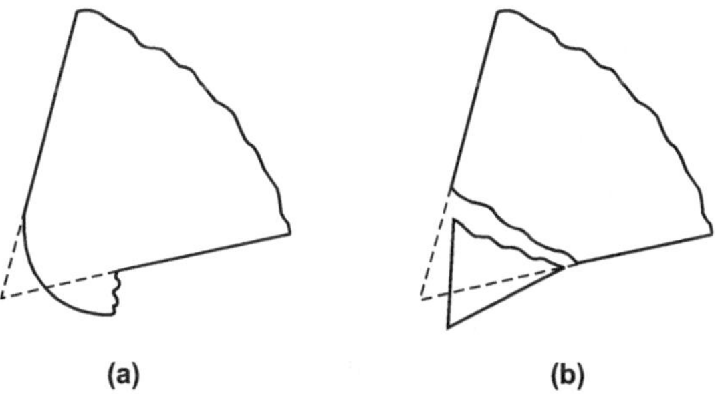

Fig. 3.14 : Mechanical breakage of tool

4. Thermal Cracking and Softening

- Due to heat generated in the metal cutting process, the tool tip and the area closer to the cutting edge becomes very hot.
- If this heat crosses the high temp, at which the tool losses its hardness the tool material starts deforming plastically.
- This is said to have failed due to softening.
- The factors responsible for this are cutting speed, high feed rate, and excessive depth of cut, smaller nose radius and the choice of a wrong tool material.

5. Diffusion Wear

- This occurs because of the diffusion of metal and carbon atoms from the tool surface into the work material and the chips.

6. Plastic Deformation

- When high compressive stresses act on the tool rake face, the tool may be deformed downwards and this deformation takes place primarily in the nose area of the insert and reduces the relief angle.

- Deformation leads to the sudden failure of the tool by fracture or localized heating.

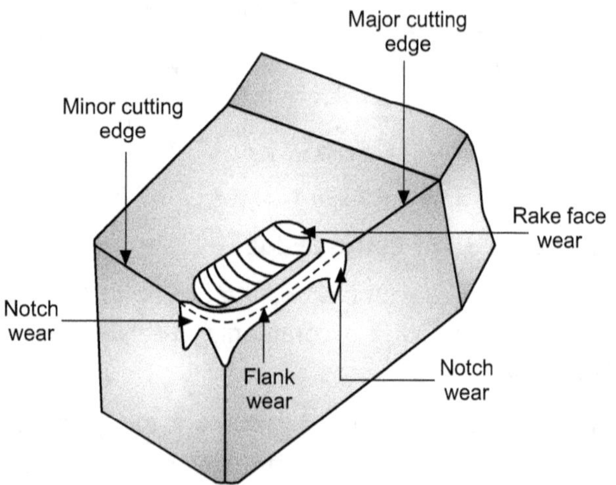

Fig. 3.15 : Types of tool wear

Factors on Tool Wear Depends

- **Cutting Speed :** More cutting speed, more tool wear.

- **Feed and Depth of Cut :** More feed and depth of cut, more is the tool wear.

- **Tool Geometry :** Tool wear also depends upon rake angle, clearance angle, cutting edge angle, nose radius.

- **Tool Material :** Hardest is the material, lower is the tool wear. Softest is the material, more is the tool wear.

- **Cutting Fluid :** More we use the cutting fluid, less will be friction and less tool wear.

3.8 CUTTING TOOL MATERIALS

The cutting tool materials must possess a number of important properties to avoid excessive wear, fracture failure and high temperatures in cutting. These are hot hardness, toughness, low friction, and wear and abrasion resistance etc. Following are important tool material.

1. Carbon Steels / High Carbon Steels

- It is the oldest of tool material.

- The carbon content is 0.6-1.5% with small quantities of silicon, chromium, manganese, and vanadium to refine grain size.

- This material has low wear resistance and low hot hardness.

- The use of these materials now is very limited.

Advantage :

- Fabrication is easy and easy to harden.

Disadvantages

- Suitable only for machining of ductile material at low cutting speeds.

- They are not able to withstand at high temperatures.

2. High-Speed Steel (HSS)

- They are highly alloyed with vanadium, cobalt, molybdenum, tungsten and chromium added to increase hot hardness and wear resistance.

- The high toughness and good wear resistance make HSS suitable for all type of cutting tools with complex shapes for relatively low to medium cutting speeds.

- The most widely used tool material today for taps, drills, reamers, gear tools, broaches, etc.

Advantage :

- High machining performance with longer tool life.

Disadvantage :

- Very costly and difficult to fabricate.

3. Cemented Carbides

- These are the most important tool materials today because of their high hot hardness and wear resistance.

- The main disadvantage of cemented carbides is their low toughness.

- In spite of more traditional tool materials, cemented carbides are available as inserts produced by powder metallurgy process.

Advantages

- Used for high cutting speeds.

- Very high hardness and wear resistance.

Disadvantages

- They are costly.

- Low toughness because of brittleness.

4. Ceramics

- Ceramic materials are composed primarily of fine-grained, high-purity aluminum oxide (Al_2O_3), pressed and sintered with no binder.
- Cold-pressed ceramic and hot-pressed ceramics(cermets) Both materials have very high wear resistance but low toughness, therefore they are suitable only for continuous operations such as finishing turning of cast iron and steel at very high speeds.
- There is no occurrence of built-up edge, and coolants are not required.

Advantages

- High compressive strength.
- High hardness
- High wear resistance and longer tool life.
- Withstand upto 1200°C temp.

Disadvantages

- Very brittle and not suitable for cutting under impact load.
- Low thermal conductivity.

5. Cubic Boron Nitride (CBN)

- CBN is the hardest tool material. CBN is used mainly as coating material because it is very brittle.
- In spite of diamond, CBN is suitable for cutting ferrous materials.

Advantages

- CBN has high hardness and high thermal conductivity.
- The life of a CBN tool is 4 to 5 times higher than that of a diamond tool.

Applications : As a grinding wheel on HSS tools for machining high temperature alloys, titanium, nimonic stainless steel, stellite and chilled CI.

6. Diamond

- The diamond is the hardest known material and can be run at cutting speeds about 50 times greater than HSS and at temperatures upto 1650°C.
- This is used when good surface finish and dimensional accuracy are desired.

Advantages

- It is very hard and has very low co-efficient of friction.
- Low thermal expansion and high heat conductivity.

Disadvantages

- It is very costly.
- Because of its harness, it is very difficult to produce the required shape (fabricate).

Applications : Diamonds are used for cutting very hard materials such as glass, ceramics, abrasives and steels .It is also used for dressing the grinding wheel.

7. Stellite

- Stellite is the trade name of a non-ferrous cast alloy composed of cobalt, chromium and tungsten.
- The range of elements in these alloys is 40-48% cobalt, 30-35% chromium, 12-19% tungsten, 1.8 to 2.5 % carbon.
- They cannot be forged to shape, but may be deposited directly on the tool shank in an oxy-acetylene flame.

Advantages

- They have wear resistance.
- Retain its hardness upto 1000°C.
- It can be used at high cutting speeds.

Disadvantage :

- It is brittle and cannot be used under impact machining conditions.

3.9 TOOL COATINGS

PVD Coatings, Cutting Tools and Workpiece Materials in Machining

- Coatings improve cutting tool performance
  - What does it mean to the customer
  - Machine shop economics
- Technical review of functional coatings
  - Understanding tool materials
  - Understanding workpiece metallurgy and machinability
  - Role of coatings in metal cutting
- New coating introduced in the market
  - How do coatings fit the application
  - Coating applications guide

What does the coating do for the machine shop ?

- Tool life increases, 30% to 200%, depending on application
  - Save on tool costs
  - Fewer tool changes and less downtime on machine.

- Higher speeds (sometimes higher feed rates) are possible.
- Dry (or near-dry) machining in certain operations
  - ➤ Environmental impact
  - ➤ Decrease coolant disposal costs

Fig. 3.16

3.10 MACHINABILITY   [Nov. 16, 17, May 17]

- The ease with which a given material may be worked with a cutting tool is known as machinability.
- In the most general case good machinability means that material is cut with good surface finish, long tool life, low force and power requirements, and low cost.

Machinability Depends on

- Chemical composition of w/p material.
- Micro structure.
- Mechanical properties.
- Physical properties.
- Cutting conditions.

3.10.1 Factors Affecting Machinability

Common machine variables affecting ease of cutting are,

- Cutting speed
- Dimensions of the cut
- Tool material
- Tool form (angles, radii, etc.,)
- Cutting fluid
- Nature of engagement of tool with the work

3.11 CUTTING FLUIDS

Cutting fluids, sometimes referred to as lubricants, coolants are liquids and gases applied to the tool and w/p to assist on the cutting operations.

3.11.1 Purpose of Cutting Fluid (Functions)

- To cool the tool.
- To cool the w/p.
- To wash the chip away from the tool.
- To improve surface finish and tool life 30%.
- To Reduces force and energy consumptions.
- To cause chips break up into small part.
- To protect the finished surface from corrosion.

3.11.2 Properties of Cutting Fluid

- Good lubricating qualities to produce low-co-efficient of friction.
- High flash point so as to eliminate the hazard of fire.
- High heat absorption for readily absorbing the heat developed.
- Harmless to the skin of the operators.
- Non-corrosive to the work or the m/c.
- Low priced to minimize production cost.
- Low viscosity to permit free flow of the liquid.
- Harmless to the bearings.
- Should be chemically natural inert.
- Transparency so that the cutting action of the tool may be observed.

3.11.3 Types of Cutting Fluids

1. Water

It is principally a coolant and not a lubricant water with alkali, salt or water soluble additive and little soap are sometimes used as a coolant. But water alone is objectionable for its corrosiveness.

2. Soluble Oils

These are emulsions composed of around 80% or more of water, soap and mineral oil. The soap acts as an emulsifying agent which breaks the oil into minute particles to disperse them throughout water. The water increases the cooling effect, and the oil provides the best lubricating properties and ensures freedom from rust.

3. Straight Oils

These are straight mineral oils (petroleum), Kerosene, low viscosity petroleum factions. Straight fixed or fatty oils consisting of animal, vegetable, lard oil etc. They have both cooling and lubricating properties and are used in light machining operations.

4. Mixed Oils

This is a combination of straight mineral and straight fatty oil. This makes an excellent lubricant and coolant for automatic screw m/c work and where accuracy and good surface finish are of prime importance.

SOLVED EXAMPLES

Example 3.1 : *The tool life with HSS tool for machining mild steel at 18 m/min is 2 hours. Calculate tool life when the tool operates at 24 m/min.*

Solution : Given, Tool life, T_1 = 2 hours = 2 × 60 = 120 min

Cutting speed, V_1 = 18 m/min

$$V_2 = 24 \text{ m/min}$$

$$T_2 = ?$$

As we know,

Taylor's tool life equation,

$$VT^n = C \begin{bmatrix} \because C - \text{constant} \\ n - \text{taylor's exponents} \end{bmatrix}$$

$$V_1 T_1^n = C \text{ for HSS } n = 0.1 \text{ to } 0.15$$

$$18 \times (120)^{0.125} = C \quad n = 0.125 \text{ (avg.)}$$

$$C = 19.11$$

Now,

for $\quad V = 24$ m/min

$$V_2 T_2^n = C$$

$$24 \times (T_2)^{0.125} = 19.11$$

$$T_2 = \left(\frac{19.11}{24}\right)^{\frac{1}{0.125}}$$

$$T_2 = 0.16 \text{ min}$$

Example 3.2 : *By using tool life equation with n = 0.5 and c = 400 calculate the percentage increase in tool life when cutting speed is reduced by 60%.*

Solution : Given, n = 0.5, c = 400

Taylor's tool life equation

$$VT^n = C$$

$$V_1 T_1^n = V_2 T_2^n$$

for 60% reduction in tool life,

$$V_2 = 0.4 \, V_1$$

$$\therefore \quad V_1 T_1^n = 0.4 \, V_1 T_2^n$$

$$\therefore \quad \left(\frac{T_2}{T_1}\right)^n = \frac{1}{0.4}$$

$$\frac{T_2}{T_1} = \left(\frac{1}{0.4}\right)^{\frac{1}{n}}$$

$$\therefore \quad \frac{T_2}{T_1} = 6.25$$

$$\therefore \quad T_2 = 6.25 \, T_1$$

$$\therefore \quad \text{Percentage increase} = \frac{T_2 - T_1}{T_1} \times 100$$

$$= \frac{6.25 \, T_1 - T_1}{T_1} \times 100$$

$$= 525 \, \%$$

$\therefore$ 525% increase in tool life when cutting speed reduced by 60%.

Example 3.3 : *The following equation for tool life is used for turning operation.*

$$V \, T^{0.125} \, f^{0.77} \, d^{0.36} = C$$

A 60 min toll life was obtained while cutting at V = 30 m/min, f = 0.35 mm/rev. and depth of cut d = 2.5 mm. Calculate change in tool life when cutting speed, feed and depth of cut are increased by 30% individually.

Solution : Given, T = 60 min, V = 30 m/min, f = 0.35 mm/rev, d = 2.5 mm

Now, $\quad V \, T^{0.125} \, f^{0.77} \, d^{0.36} = C$

$$30 \times (60)^{0.125} \times (0.35)^{0.77} \times (2.5)^{0.36} = C$$

$$\therefore \quad C = 31.015$$

(i) If cutting speed is increased by 30% and remaining parameter are same.

then,

$$T^{0.125} = \frac{31.015}{39 \times (0.35)^{0.77} \times (2.5)^{0.36}}$$

and $\quad V = 30 \times 1.3 = 39$ m/min

$$T^{0.125} = 1.283 \Rightarrow T = 7.36 \text{ min}$$

(ii) If feed is increased by 30% i.e.

$$f = 0.35 \times 1.3 = 0.455 \text{ mm/rev.}$$

then,

$$T^{0.125} = \frac{31.015}{36 \times (0.455)^{0.77} \times (2.5)^{0.36}}$$

$$T^{0.125} = 1.136 \Rightarrow T = 2.77 \text{ min}$$

(iii) If depth of cut is increased by 30%

i.e. d = 2.5 × 1.3 = 3.25 mm

$$T^{0.125} = \frac{31.015}{36 \times (0.35)^{0.77} \times (3.25)^{0.36}}$$

$$\Rightarrow \quad T^{0.125} = 1.264 \Rightarrow T = 6.55 \text{ min.}$$

EXERCISE

1. Under what conditions are diamond, boron carbide and cubic boron nitride used as abrasive materials for making grinding wheels?

2. What is meant by 'grain size' of an abrasive material?

3. What is meant by dressing and truing of grinding wheels?

4. Define "grinding ratio".

5. Sketch and explain the working of an external cylindrical grinding machine.

6. Sketch and explain the three methods of cylindrical grinding.

7. What is the difference between plain and universal cylindrical grinders?

8. Describe in detail how an internal grinder operates.

9. What are the advantages and disadvantages of centreless grinding ?

10. Sketch and explain the three methods of external cylindrical centreless grinding.

PROCESSING OF POWDER METALS

4.1 INTRODUCTION [Dec. 11]

- Powder metallurgy may be defined as the process of manufacturing components from metal powders by compaction and sintering. Many metal powders and their oxides are used in cosmetics, paintings and for decorative purposes for the last few centuries. Some of the components manufactured by powder metallurgical technique have unique applications, which may not be obtained by conventional processes.

- Powder metallurgy (P/M) deals with,
  - ➢ Powder manufacturing
  - ➢ Compaction and
  - ➢ Sintering.

- Powder metallurgical parts find applications in the various areas such as
  - ➢ Automobiles
  - ➢ Aerospace
  - ➢ High temperature parts
  - ➢ Nuclear reactor parts
  - ➢ Superconducting materials and
  - ➢ Tool materials.

- During the production of powder metallurgical parts, porosity cannot be avoided. However, this porosity becomes a unique property and is used in few applications like self-lubricating bearings. The final properties of the component depend upon properties of the metal powder as size, shape and sintering properties. Similarly, refractory materials like tungsten, molybdenum, zirconium etc. can be processed only by powder metallurgy technique. The applications of powder metallurgical parts are explained in detail at the end of this chapter.

4.2 PRODUCTION OF METAL POWDERS

- Depending upon the final application of the powder, various methods of powder production are used. Each method produces powder with specific characteristics.

- These methods are classified as follows :

1. **Mechanical Processes**
- Machining
- Crushing
- Milling

- Shotting
- Graining and
- Atomization

2. **Physico-Chemical Processes**
- Condensation
- Thermal decomposition
- Reduction
- Electrodeposition
- Precipitation from aqueous solution
- Precipitation from fused salt
- Hydrometallurgical reduction
- Intergranular corrosion and
- Oxidation and decarburisation

4.2.1 Mechanical Processes (May 10; Dec. 10, 13)

1. **Machining**
- This method is mainly used to produce fillings, turnings, chips, etc. These can be pulverized by crushing and milling. Very coarse and bulky powders are obtained by this process.

- In this process, irregular shaped particles are produced. This is used for the production of magnesium powder (for pyrotechnic applications), beryllium powders, silver solders and dental alloys.

2. **Crushing**
- This method is used for disintegration of oxides and brittle materials. Various crushing instruments such as stamps, hammers, jaw crushers etc. are used. The powder produced by this method is of angular shape for brittle material and of flaky shape for ductile materials.

- Titanium, zirconium, vanadium etc. can be powdered by this method. Some of the metals and alloys can be hardened first and then crushed for making powders.

3. **Milling**
- This is one of the most useful method with which various fine grades of powders can be produced. Milling or grinding can be done by using ball mill, rod mill, impact mill, disk mill, vortex mill etc.

- In ball milling, the material to be disintegrated is tumbled in a container with a large number of hard wear resistant solid balls. These balls hit the material and break it. The speed and time of rotation of the container decides the nature of final product. The milling is carried out either in wet or dry condition. The ball mill container is of stainless steel or steel lined with hard alloy steel plates. The balls used are of white cast iron or hardened steel.

- The major disadvantages of milling are work hardening, excessive oxidation of the final powder, particle welding and agglomeration. Annealing in a reducing atmosphere eliminates these problems.

- This method is widely employed for carbide-metal mixture and cement materials to perform blending and particle size reduction.

4. Shotting

- The method consists of pouring a fine stream of molten metal through a vibrating screen into air or neutral atmosphere like nitrogen.

- The molten metal gets disintegrated into a large number of fine droplets, which solidify as spherical particles.

- The opening of screen, its frequency for vibration and the melt temperature decide the properties of the powder produced. This method is mostly used for non-ferrous metals.

5. Graining

- This method uses the same principle as above, with the only difference that the solidification is allowed to take place in water. Finally, pulverisation is used to produce fine powders.

- The metals like zinc, bismuth, tin etc. are pulverised by this method.

6. Atomization   (Dec. 13; May 14)

- The principle of this method consists of disintegration of a stream of molten metal into the fine particles mechanically by using a jet of compressed air, inert gases or water (See Fig. 4.1).

- Because of its various advantages, this method has wide applications. This method is mostly used for powdering of the metals such as tin, lead, zinc and aluminium, which have low melting points.

- Atomized products are generally sphere-shaped particles.

- By varying temperature of the metal, pressure and temperature of the atomizing gas, rate of flow of metal through orifice and nozzle decide the particle size, shape and distribution.

- It is a very flexible method, but not suitable for the refractory metals.

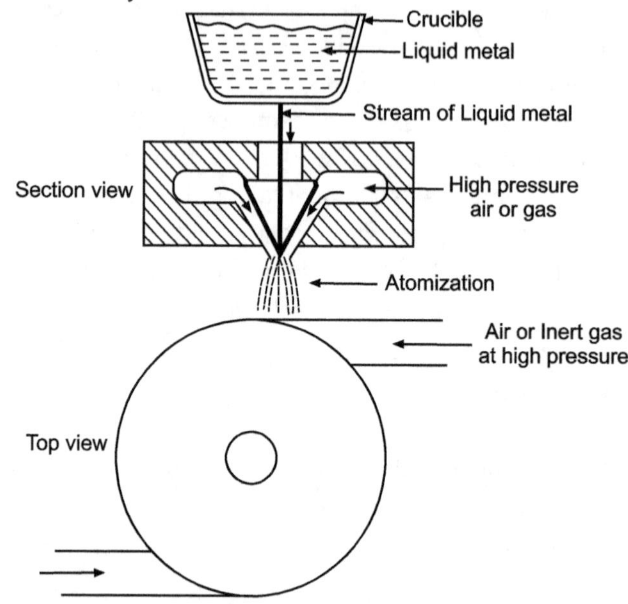

Fig. 4.1: Atomization for powder manufacturing

4.2.2 Physico-Chemical Processes

1. Condensation

- This is a modification of usual distillation process. In this method, the volatile metal is distilled off and the vapours are condensed to produce powders (See Fig. 4.2). Metals like zinc, magnesium etc. are used to produce their respective powders. Zinc dust, at least 97% pure is produced in a very large quantity of tonnes by this way.

- Powders with very small particle size, less than 500 A° in diameter can be produced by this method.

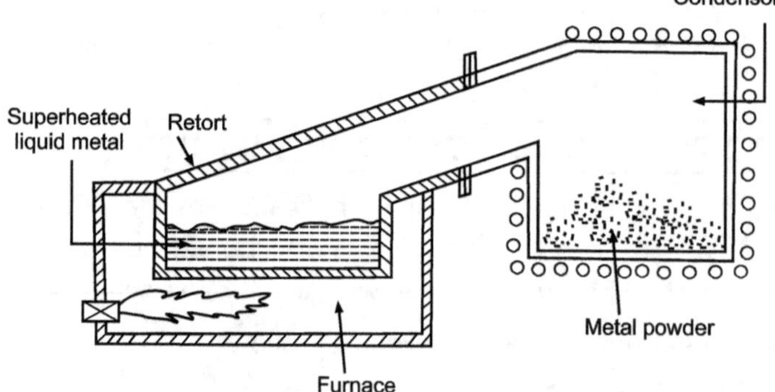

Fig. 4.2: Metal powder manufacture by condensation method

2. Thermal Decomposition

- This method, termed as gaseous pyrolysis method is used for powdering of the metals like iron, nickel, zinc, magnesium, cobalt, chromium etc. The metal carbonyls are volatile liquids. They can be easily decomposed by forming powders.

e.g. Iron pentacarbonyls $Fe(CO)_5$ can be thermally decomposed to form iron powder.

$$Fe(CO)_5 \rightarrow Fe + 5\,CO$$

Similarly, nickel tetracarbonyl $Ni(CO)_4$ gives nickel powder.

$$Ni(CO)_4 \rightarrow Ni + 4\,CO$$

- These carbonyls may be produced by passing CO over a spongy or powdered metal at some suitable temperature and pressure. This process gives very fine powders. It is also economical as carbon monoxide can be recycled.

- Iron powders produced by this method are perfectly spherical and purest of all the commercial metals (over 99.5%). The iron-nickel powder produced by decomposition of their carbonyls is used for the manufacture of high permeability cores for long distance communication apparatus. Tungsten powder produced by this method is used for welding tungsten filaments.

3. Reduction Method

- In this method, the metal powder is produced by reducing oxides or halides of metals, using a suitable reducing agent. This is one of the economical and flexible methods. It is extensively used for the manufacture of iron, copper, nickel, tungsten, cobalt powders.

- The process produces extremely fine and irregularly shaped particles with a considerable porosity. Normally carbon, hydrogen, ammonia and carbon monoxide are used as reducing agents.

- Large scale production of iron powders at low cost is possible by mixing high grade oxides with carbon and heating them in a ceramic container, because of low cost and availability of raw material. Metallic reducing agents are also used to reduce oxides e.g. chromium powder is produced by reduction of Cr_2O_3 with magnesium; zirconium powder by reduction of zirconium oxide with calcium and so on. Since, oxides are very brittle, fine powders are possible to be produced.

- W and Mo powders are prepared from WO_3 and MoO_2 by reduction with H_2 for their use in manufacture of incandescent lamp filaments, radiovalves, x-ray target etc. In most of the cases, irregular shaped fine powders are produced.

4. Electrodeposition

- In this method, metal powders can be produced by electrodecomposition from aqueous solutions and fused salts.

- This method is a reverse of electroplating. In electroplating process, the parameters are adjusted to obtain adherent and continuous deposition, while in electro-deposition, coarse, loose and non-adherent deposition is required to recollect the powder.

- This technique is mainly employed for the commercial production of metal powders from copper, beryllium, iron, zinc, tin, nickel etc.

- Following three types of electro deposition are practically used

  (i) Deposition as a hard and brittle mass, which may be ground to form powders;

  (ii) Deposition as a soft, spongy mass, which may be powdered by light rubbing and

  (iii) Direct deposition in powder form from the electrolyte, which drops to the bottom of the cell.

- The last two methods are used for production on the commercial line. However, the process is not suitable for production of alloy powders.

- Following conditions of electrolyte favour the powder manufacture

  ➢ High current density,

  ➢ Low metal-ion concentration,

  ➢ High acidity,

  ➢ Low temperature,

  ➢ High viscosity and

  ➢ Circulation of electrolyte

- The powders produced by this process are crystalline and dendritic having less flow rate in some cases. The purity as high as 99.9% in the case of non-ferrous metals may be reached.

5. Precipitation from Aqueous Solution

- This process is based on the principle that less noble metal displaces more noble metal in the electromotive series, from an aqueous solution.

- This process produces very fine powders. For example, silver powder is produced from its nitrate solution by adding copper or iron; tin powder is precipitated by metallic zinc from stannous chloride solution.

- A very common method of producing copper powder is by precipitation from sulphate solution with iron.

- The precipitated metal powders are porous, but with excellent purity. This process produces dendritic shaped powder.

6. Precipitation from Fused Salts

- Reactive metal powders are produced by precipitation from fused salts.

- For example, $ZrCl_4$ salt is mixed with an equal amount of KCl and some magnesium to produce Zr powder. Similarly, beryllium and thorium powders are produced.

7. Hydrometallurgical Reduction

- Nickel, cobalt and copper powders are mainly produced by this method.

- The reduction of aqueous solutions or slurries of salts of these metals are made with hydrogen under controlled conditions of pressure and temperature.

$$M^{++} + H_2 \rightarrow M + 2H^+$$

- All metals must be washed to free from the traces of salts and then dried. This method is also used to produce composite powders.

- The process produces pure metal powders of spherical shape.

8. Intergranular Corrosion

- Usually, grain boundaries of any metal corrode very fast than the grains. This is intergranular corrosion. In this method, the grain boundary area is intentionally corroded by suitable electrolyte. This process is used to produce stainless steel powder.

- Stainless steel is a type of alloy steel containing chromium and nickel. When stainless steel (particularly austenitic stainless steel) is heated in the temperature range of 500 to 800°C for a long time, the chromium combines with carbon and forms chromium carbides. These complex carbides get precipitated at grain boundary. The chromium, which is added to improve the corrosion resistance, thus, gets consumed with carbon and the corrosion resistance is drastically dropped. Such steel becomes sensitized in corrosive medium and intergranular corrosion i.e. separation of grain occurs.

- This network of carbide is preferentially corroded by boiling aqueous solution of 11% copper sulphate and 10% sulphuric acid. Rapid disintegration may be obtained using stainless steel as anode and solutions of copper sulphate and sulphuric acid as an electrolyte in a cell.

- These powder particles possess angular shape and the final particle size depends on the size of initial grain structure.

- This method was used in the past. However, atomization is more suitable now-a-days in view of mass production.

9. Oxidation and Decarburisation

- This method is mainly used for the production of pure reactive metal powders, particularly niobium.

- The process consists of reacting metal carbide with metal oxide in vacuum at a higher temperature so that the oxygen and carbon are removed as CO and finally the metal remains in powder form.

- However, due to its complexity, the process is not so useful on commercial footing.

4.3 CHARACTERISTICS AND TESTING OF METAL POWDERS

- The method of production of metal powder decides its properties. The mechanical behaviour of the powder and its metallurgical component depends upon the characteristics of initial metal powder. So, it becomes necessary to test various properties of metal powders.

- The main purpose of powder testing is to ensure, whether or not the powder is suitable for further processing.

- The principal characteristics of a metal powder are,
  - ➢ Chemical composition and purity
  - ➢ Particle size and its distribution
  - ➢ Particle shape
  - ➢ Particle porosity
  - ➢ Particle microstructure.

- **Other Characteristics Include**
  - ➢ Specific surface
  - ➢ Apparent density
  - ➢ Tap density
  - ➢ Flow rate
  - ➢ Compacting and sintering properties.

Sampling:

- The sample used must be a true representative of the entire batch produced. There are various methods used for making samples.

- Coning and quartering method involves pouring of powder on a polished metal sheet and splitting up into four equal parts. This process is repeated till the required sample quantity is obtained.

- The scoop sampling method consists in inserting a scoop into a thoroughly mixed powder in a container and then drawing it full of powder as a sample. Any sampling method can be adopted. It can be standardised as per individual's experience.

1. Chemical Composition

- The chemical composition of any powder reveals the type and percentage of impurities it contains and determines particle hardness and pressing properties. Impurity refers to undesirable metal contents, which affect the mechanical properties of the component.

- Insoluble oxides having more hardness than that of metal powder, increase abrasion of die and punch. Non-uniform pressing and sintering occurs due to these insoluble oxides.

- The chemical composition of powder is determined by the well established techniques of chemical analysis. This includes,

  - ➢ Gravimetric analysis,
  - ➢ Volumetric analysis,
  - ➢ Electrochemical analysis,
  - ➢ Colorimetric analysis etc.

- Oxygen content is determined by wet analysis. Some of the insoluble contents may be determined by dissolving the powder followed by filtering, igniting and weighing the residue.

2. Particle Size

- Particle size affects mould strength, density of compact, porosity, dimensional stability etc. Particle size is expressed by the diameter of the spherical shaped particles and by the average diameter for non-spherical particles.

- For all practical purposes, the selection of powder size for a specific application is usually based upon one's own experience. However, no doubt fine powder is always preferred to coarse one.

- Metal powders are divided into three categories

  - ➢ Sieve,
  - ➢ Sub-sieve and
  - ➢ Sub-micron or ultrafine.

- A majority of metal powders employed in powder metallurgical industries vary in size from 4 to 250 microns. Very fine powders tend to be pyrophoric i.e. those which oxidise quickly in the atmosphere.

(i) Sieve Method

- This is the most popular and simple method. Different sieves are used for classifying powders. Standard sieves of different mesh sizes are used. This mesh size is standardised by the various international organisations.

- The opening of sieve is expressed by the number of meshes per linear inch. Woven wire sieves are made of copper, brass, nickel or even of nylon.

- Different mesh sieves are stacked over each other. The coarse sieve is at the top. The fineness goes on increasing towards the bottom.

- A sample of 100 gm metal powder placed on the top sieve is shaken or vibrated to provide two motions viz. circular and translating, for 15 minutes. The quantity of powder retained in each sieve is taken out and weighed accurately from these weights, size and size distribution can be found out.

- The particles, which pass through the sieve, are given minus number and the particles which are retained on the sieve, are given plus numbers of that sieve. The size distribution is expressed by weight fraction of powder retained on each sieve.

- The powders within the range of 44 to 840 microns can be successfully sieved.

(ii) Microscopic Method

- Microscopic sizing or counting method depends largely on the skill. But, this is a direct method.

- It involves actual counting of particles and individual examination of a large number of particles on a glass slide. This is a time taking process, but one of the most reliable methods. Optical and electron microscopes are used for this measurement.

- As the powder is always of mixed sizes, a number of readings should be taken and an average may be mentioned for its particle size. Shape of the particles can also be visualised very easily.

- Optical microscope is used for the determination of particle diameters down to about 0.3 microns. An electron microscope is used for the particles in the diameter range of 10 to 0.001 micron.

(iii) Sedimentation Method

- This is a method of classifying metal powders according to settling velocities in a fluid.

- Sedimentation involves suspending the powder sample by using a proper stirring in a fluid medium and allowing it to settle for a suitable time. The settling velocity in suspension is measured. The settling velocity of a spherical particle is proportional to square of the particle diameter.

- So particles having equal sizes can be separated as their settling time will be the same. The amount of particles settling at different intervals of time is measured. Irregularly shaped particle size is defined as the diameter of a sphere of the same material having the same settling velocity under constant conditions.

- Laboratory methods for sizing by sedimentation may be divided into two groups

  - ➤ The fractionating methods are use, when it is essential to separate individual size fractions for individual examinations.

  - ➤ Non-fractionating methods are used for determining size distribution and estimating specific surface.

- Following are some of the methods, which use the principle of sedimentation

  - ➤ Hydrometer or specific gravity method uses change in the specific gravity of suspension at various levels of depths.

  - ➤ Manometric method measures the change in the density of suspension by the change in the pressure at various levels of height.

  - ➤ There is one method called turbidimetric method based on change in intensity of transmitted light through dilute suspension at certain depth. However, the selection of test method depends on the various working parameters.

(iv) Elutriation Method

- This method is used for sub-sieve powders.

- In this method, the metal powder is allowed to settle in a moving liquid or gas of constant velocity. The particles with settling velocity less than the velocity of rising liquid or gas are carried upwards, while those with higher settling rate fall in the column. By changing the velocity of the medium, the particles can be separated according to their sizes.

- This is a fractioning method and is used for the separation as well as determination of size fractions of the powders. This method is also useful to remove unwanted fine materials from the powder.

- Elutriation in air is widely used for metal powders.

3. Particle Shape

Various shapes of metal powders are observed according to the method of their production (See Fig. 4.3), e.g.

- **Spherical Shape:** Carbonyl iron, condensed zinc, lead, atomization, precipitation from aqueous solution by gases.

- **Rounded or Droplets:** Atomized copper, zinc, aluminium, tin, chemical decomposition.

- **Angular Shape:** Mechanically disintegrated antimony, cast iron, stainless steel by process of intergranular corrosion.

- **Accicular Shape:** Chemical decomposition.

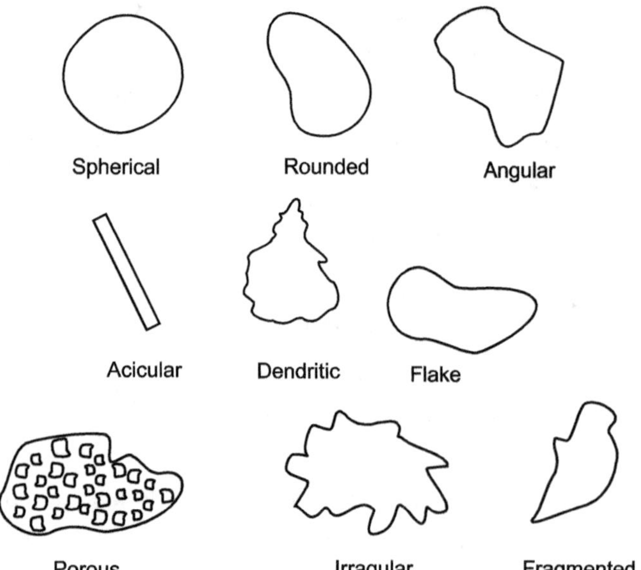

Fig. 4.3: Different shapes of metal powders

- **Dendritic Shape:** Electrolytic silver, copper.

- **Flake Shape:** Ball milled copper, aluminium.

- **Porous:** Reduction of oxides.

- **Irregular Shape:** Atomization, reduction.

- The packing capacity of powder depends on its shape. Irregular shaped particles give less density and flow rate with good pressing and sintering properties. Spherical particles have maximum density, flow rate and good sintering properties, but reduced pressing properties. Similarly, dendritic shaped powder shows less density and flow rate.

- For particle shape evaluation, optical or electron microscopic examination method is used. This is a direct method. It is also expressed by shape factor, which is the ratio of the length of particle to its breadth or surface area to size of the particle.

4. Particle Porosity

- The porosity of particles affects impact and fatigue strength as well as tensile properties of component. Porosity acts as a stress raiser, which is similar to nature of graphite in cast irons. It is practically impossible to produce P/M components without porosity. In fact, porosity itself becomes a necessary property in the case of self-lubricating bearings.

- The porosity may be present as,
  - ➤ Interconnected or
  - ➤ Separate pores.

- Porosity may be measured by mercury porosimetry (for interconnected pores). It consists of keeping P/M component with mercury in vacuum at the room temperature. The pressure of mercury is increased and the volume of penetrating mercury and density of porous material is measured.

- Porous parts can be studied similar to a metal sample (i.e. by following mounting, polishing and without etching).

5. Particle Microstructure

- For observation of microstructure, the same method of metallography i.e. polishing and etching is used. Smearing of surface layers during polishing may cover up the pores. So, extra precautions should be taken. After etching, the specimen should be thoroughly washed to avoid absorption of etchant in pores. So, electrolytic etching may be used.

- Metallographic examination reveals amount and distribution of porosity, phases, grain size, inclusion and heterogeneity in structure.

4.4 OTHER CHARACTERISTIC OF METAL POWDERS

1. Specific Surface

- The specific surface of powder may be defined as the total surface area of a particle per unit weight. The specific surface depends on size, shape, density and surface conditions of the particle.

- The specific surface i.e. contact area between powder particles strongly affects compacting and sintering properties. High surface area increases sintering rate as well as entrapment of air. This shows bridging effect, which causes cracks.

- Coarser powder with smaller contact area gives bad sintering and weak mechanical properties.

- Specific surface may be measured by,
  - ➤ Adsorption method and
  - ➤ Permeability method.

- The permeability method is easy and quick. It may be used to control the process. This method involves measurement of pressure drop across the bed of packed powder particles contained in a chamber with relation to fluid flow. The surface area of powder can be calculated by measuring the resistance of a packed column of the powder to the flow of liquid.

- The specific surface can be calculated from the size distribution data either by graphical method or by numerical calculation.

- Typical values of specific surface of iron powder are as follows

Powder	Particle Size (μ)	Specific Surface (cm²/gm)
Iron	65	510
	50	950
	6	5200

2. Apparent Density

- Apparent density is called as packing density or loading weight. It is defined as mass per unit volume of loose or unpacked metal powder. It includes internal pores, but excludes external pores. It depends on :
  - ➤ Chemical composition
  - ➤ Particle shape, size, distribution and
  - ➤ Method of manufacture.

- The apparent density decreases with decrease in size of particle and increase in the surface roughness. Coarse and spherical shaped powder shows good apparent density, while irregular and flake shaped powder shows poor apparent density. Very fine powder shows poor apparent density.

- It strongly affects pressing and sintering properties. Longer compression strokes are required for lower apparent density powders.

3. Tap Density

- Tap density or load factor is the apparent density of the powder after it is mechanically vibrated or tapped until the level of the powder no longer falls. This becomes useful for storage, packing or transport of commercial powders. This shows a similar effect as that of apparent density on pressing and sintering properties. Tap density increases with reduced particle porosity.

4. Flow Rate

- This is one of the most important characteristics of powders. It measures the ability of powder to be transferred.

- It is defined as the rate at which metal powder flows under gravity from a container (called flow meter) through an orifice having specific size, shape and finish (See Fig. 4.4).
  Poor flow properties of powder result in slow and uneconomical feeding of a die.

- It is affected by particle size and distribution, absorption of moisture and coefficient of interparticle friction. Spherical powder shows maximum flow rate, while irregular and dendritic powders show poor flow rate. Hall flow-meter is used to measure the flow rate.

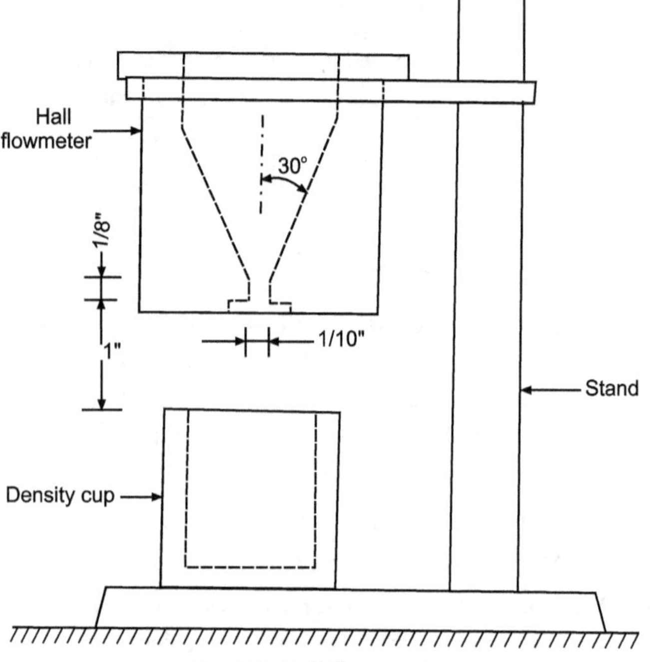

Fig. 4.4: Hall flow-meter

- It is accurately machined conical funnel made of brass with smooth finish having internal angle of 60°. The orifice situated at the bottom is either 1/8" (for ferrous powder) or 1/10" in diameter (for non-ferrous powder) and with length 1/8". The time required to flow the weighed sample of powder (around 50 gm) is measured and reported as flow rate in seconds or in gms/min in the case of non-standard weight of powder.

- Apparent density and flow properties are closely related.

5. Compacting and Sintering Properties

- These are represented by the following terms
  - ➤ Compressibility and
  - ➤ Compactibility.

- **Compressibility** affects the density of P/M parts.

- It can be defined as the measure of ability of powder to get deformed under applied load. It can also be defined as, The ratio of green density of compact to the apparent density of powder (See Fig. 4.5) or

- The ratio of height of uncompacted powder in the die to the height of the pressed compact or

- The ratio of volume of the powder poured into the die to the volume of pressed compact.

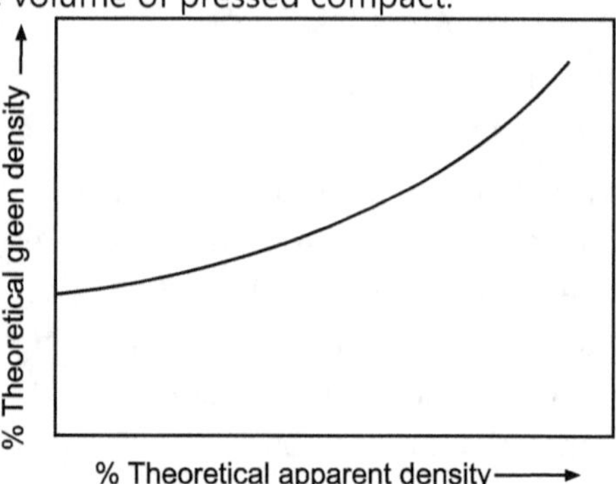

Fig. 4.5: Dependence of green density on apparent density of a metal powder

- This ratio is termed as compression ratio.
  Maximum compression ratio

$$= \frac{\text{True density of bulk material}}{\text{Apparent density}} = 2 \text{ to } 8$$

- Higher compression ratio increases friction between powder and the die wall.

- **Compactibility** is defined as the minimum pressure required to produce a powder compact of desired green strength.

- Both the above terms depend on
  - ➤ Particle size
  - ➤ Particle shape
  - ➤ Porosity
  - ➤ Hardness
  - ➤ Surface properties of powders

6. Green Density

- This is the density of a green compact i.e. density of compacted product; before sintering. It increases with

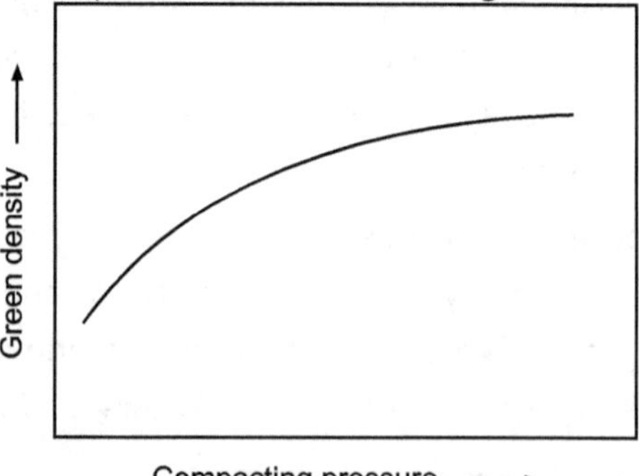

Fig. 4.6: Dependence of green density on compacting pressure

- ➤ Increase in the compaction pressure (See Fig. 4.6),
- ➤ Increase in the particle size,
- ➤ Less particle hardness and
- ➤ Low compaction speed.
- In the case of standard and regular shaped compacts (i.e. cubic, cylindrical), the green density can be calculated by its weight and dimensions. For irregular shaped compact, it is measured by weighing the sample in air and water. It is represented as gm/cc.

7. Green Strength

- It is referred as the strength of a green compact. It depends on compacting pressure. A P/M part gains green strength because of cold welding and mechanical interlocking of particles during compaction.
- It is defined as the mechanical strength required for handling a green compact without any damage to its size and shape. It is expressed in kg/mm^2.
- It depends on the size, shape, distribution, hardness etc. of powder. After compaction, the P/M parts are ejected from the die and transferred to a sintering furnace easily. So, green strength plays an important role.

The green strength may be tested by :
- ➤ Three point transverse rupture test or by
- ➤ Radial crushing test.

8. Green Spring

- After ejection of compacts from the die, expansion of compact takes place. This expansion in size is both radial and longitudinal.
- The difference between the size of compact and the die is referred as green spring.
- As the compact is ejected, elastic recovery of particles takes place. When it exceeds the green strength of the compact, cracking occurs during ejection. During manufacturing of P/M parts with close dimensional tolerance, it is necessary to determine green spring.
- In many cases, the green spring is observed to be 0.2% on the diameter and 0.5% on the length.

The green spring depends upon :
- ➤ Powder properties,
- ➤ Pressure of compaction,
- ➤ Elastic recovery of tools and
- ➤ Design of tools i.e. die and punch.

9. Sintering Properties

- **Dimensional Changes During Sintering:** Dimensional changes are observed, when sintering of a green compact is carried out. It is expressed as shrinkage.

$$\text{Percentage shrinkage} = \frac{\text{Change in length}}{\text{Sintered length}} \times 100$$

- This is used in carbide industries, where dimensional change is of the order of 25%.
- In the engineering component and bearing industries, the following formula is used for the calculation of shrinkage.

$$\text{Percentage shrinkage} = \frac{\text{Change in length}}{\text{Unsintered length}} \times 100$$

- **Sintered Density:** The method used to measure green density is used to measure sintered density also. It is related to the porosity of the finished products.
- **Porosity:** The total porosity present in sintered part can be calculated as,

$$p = 1 - \rho_p/\rho_s$$

where, p = Fractional porosity,
  ρ_p = Density of sintered part,
  ρ_s = Density of solid material.

- It is not possible to produce P/M parts without porosity remaining after sintering. The porosity badly affects the mechanical strength.
- The micro-structural observation of a sintered part follows the same metallographic techniques.

4.5 POWDER CONDITIONING

(Dec. 12, 14; May 11)

- The metal powders after production may not possess the required physical or chemical properties. Hence, powder conditioning is required, which involves mechanical, thermal or chemical treatments.

Powder Conditioning can be Explained as Below :

Heat Treatment

- Annealing heat treatment is used before mixing or blending. Annealing is carried out in reducing atmosphere or in vacuum. Annealing alters the mechanical properties.
- It eliminates Work hardening and Impuring content, also alters apparent density. High temperature heat treatment increases apparent density. This alteration in properties improves the sintering qualities.
- Annealing process forms spongy mass of metal powder and hence, pulverising and screening has to be carried out. If the powder has to be mixed with alloying elements, it is important to use the powder immediately after the annealing treatment.

Blending or Mixing

- It is an operation of thorough mixing of different powders of the same composition or of different compositions. The aim of blending is to obtain homogeneous mixture of powder.

- This helps to improve compacting and sintering properties. Additives such as binders or lubricants are usually included in these operations to control strength and porosities.

- It is very difficult to decide optimum mixing time. It must produce a homogeneous mixture in the least time.

- Various types of blenders and mills are employed for blending and mixing. Ball mills or rod mills are commonly used for mixing hard metals, but are less useful for soft metals.

- In general, double cone mixers or Y-mixers are used (See Fig. 4.7 (a) and (b)). Mixing may be either dry or wet. Wet mixing is used to produce more fine mixtures of powder particles. After this process, the powders are screened to remove unwanted material.

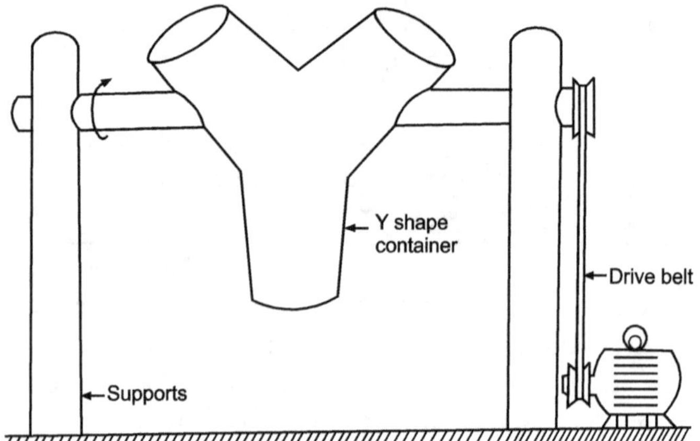

Fig. 4.7 (a): Y - cone blender

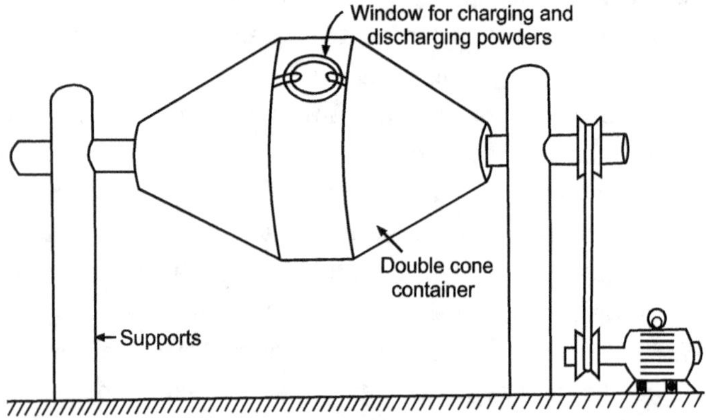

Fig. 4.7 (b): Double cone powder blender (horizontal type)

4.6 COMPACTION OF METAL POWDERS

(Dec. 10, 11, 12)

- Powder compaction is the process of forming metal powder compacts of the desired shape. Following are the various methods of powder compaction :

  - ➤ Pressureless shaping,
  - ➤ Cold pressure shaping and
  - ➤ Pressure shaping with heat.

- The aim of compacting is to consolidate the powder into desired shape, close to final dimensions. It also decides the porosity level.

- The pressure techniques involve die, isostatic, high energy rate forming, forging, extrusion, vibratory and continuous forming, while pressureless techniques involve slip casting, gravity and continuous forming.

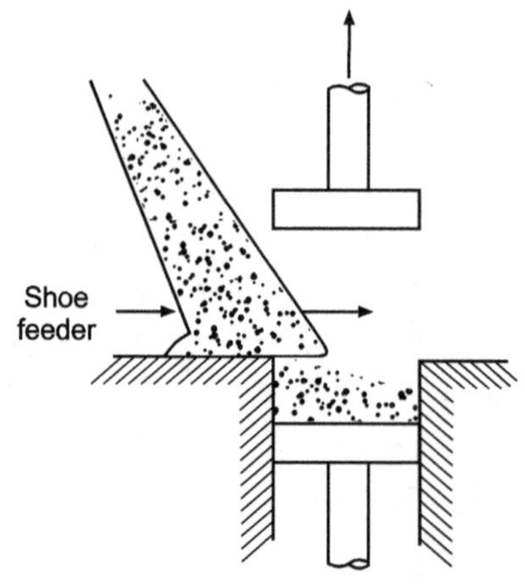

(a) Powder feeding started

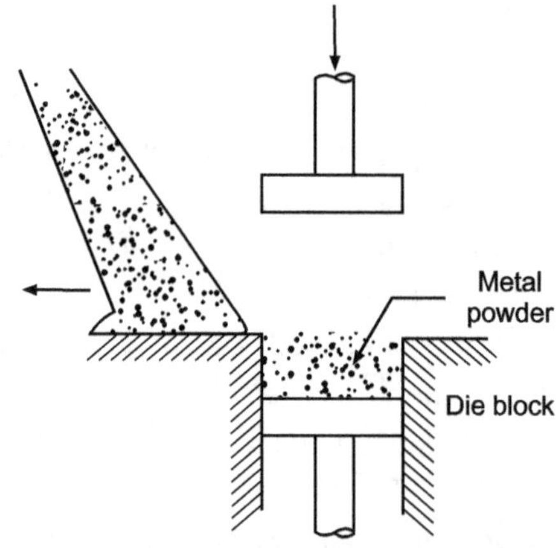

(b) Powder feeding completed

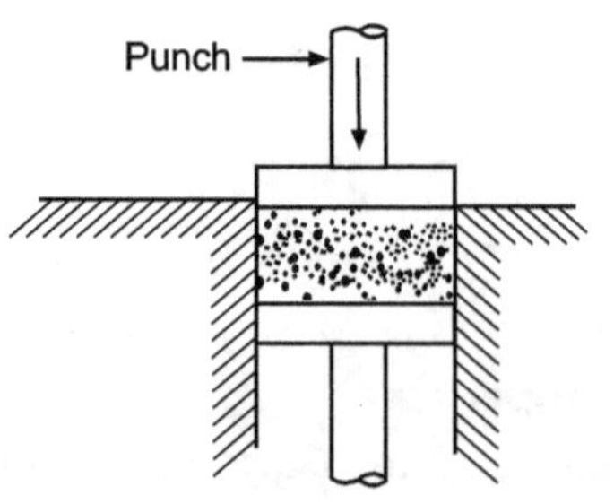

(c) Compaction

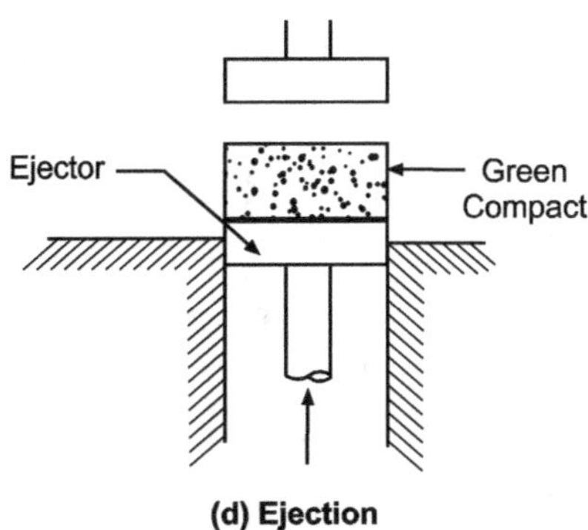

(d) Ejection

Fig. 4.8: Steps in cold compaction of metal powders

- Die compaction is a general method used in industries. It consists of the following steps :

  ➢ Filling the die cavity with a definite volume of powder.

  ➢ Application of required pressure by movement of upper and lower punches towards each other and

  ➢ Ejection of green compact by lower punch (See Fig. 4.8).

- The loads used depend on cross-section of compact and required properties :

  The pressure may be obtained by using,

  ➢ Mechanical or

  ➢ Hydraulic press.

- Mechanical presses have high speed production rates, flexible design and economical operation. Hydraulic presses have higher pressure rating, but slow stroke speeds.

- The dies are usually made of hardened, ground and lapped tool steels. The punches are made of heat treated die steels, having less hardness than the die. The alignment of punches is very important.

- The isostatic compacting consists of application of equal and simultaneous pressure from all sides. A rubber mould immersed in a fluid bath within a pressure vessel is used. A uniform green density and other properties are obtained because of uniform application of pressure. This is used for ceramic materials.

- High energy rate process gives high pressure within a short time by using some sort of explosives.

- Forging and extrusion techniques are used for limited applications. These give high density compacts. Vibratory compaction uses pressure and vibration

simultaneously. A very low pressure can give better densification in this case of compaction.

- Continuous compaction is used for simple shapes as rods, bars, tubes etc. Slip casting is also a familiar process used for ceramic powders. These are some of the compaction processes used in industries.

- Compaction pressure affects the various factors related to powders.

- Green density increases with :

  ➢ Increase in compaction pressure,

  ➢ Increase in particle size and

  ➢ Decrease in hardness.

- Better compacting technique :

  ➢ Reduces voids,

  ➢ Produces adhesion and cold welding of powders,

  ➢ Plastically deforms powder and

  ➢ Increases contact area between powder and hence, density.

4.7 SINTERING   (May 10, 13; Dec. 10, 11, 12, 13)

- This process is used to improve strength of green compact. Sintering is one of the essential last step in the manufacture of P/M part.

- It consists of heating the green compact at a temperature below the highest melting constituent. In some cases, the temperature is high enough to form some liquid phase,

  e.g. in the manufacture of cemented carbides. In this case, the sintering is done above the melting point of the binder material.

- During sintering, only surface diffusion of particles takes place. This reduces porosity and improves hardness and strength of green compact.

- Sintering may be explained by three steps as follows :

  ➢ Surface diffusion of particles

  ➢ Densification and

  ➢ Recrystallization and grain growth.

  Sintering is carried out in furnaces as

  ➢ Electric resistance furnace,

  ➢ Gas fired furnace or

  ➢ Oil fired furnace.

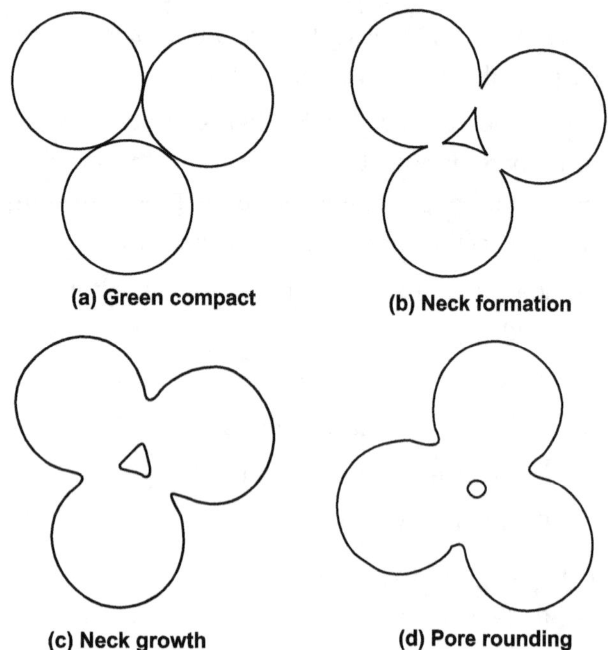

Fig. 4.9: Steps in sintering process

- Close control on temperature and atmosphere is necessary to achieve the desired properties. The formation of undesired surface films (as oxides) must be avoided as it affects the bonding between particles. So protective atmosphere should be used. The atmosphere should not contain free oxygen. It should be neutral or reducing.

- For sintering of refractory carbides and electrical contacts, a dry hydrogen atmosphere is used. Usually, for most of the commercial sintering operations, partial combustion of various hydrocarbons is used (e.g. natural gas or propane).

- The fundamental aspects of sintering involve :

  ➢ Surface contacts,

  ➢ Neck formation and growth and

  ➢ Pore rounding (See Fig. 4.9).

Elevated Levels of Temperature Favour Sintering Process

- The sintering process starts with bonding between particles as the material heats up. Bonding consists of diffusion of atoms, where, there is intimate contact between adjacent particles.

- This leads to the formation of grain boundaries. This stage results in increase in the strength and hardness. The newly formed bond areas are called as *necks*. With time and at higher temperature, neck growth takes place followed by pore rounding. The last step in the

sintering involves pore shrinkage and in rare cases, pore elimination.

- Depending on the temperature of sintering, it is classified as :

  ➢ **Solid Phase Sintering:** In this process, the compact is heated above the recrystallization temperature of low melting metal powder and

  ➢ **Liquid Phase Sintering:** In this process, the green compact is heated above the melting point of one of the alloy elements.

4.8 HOT PRESSING

- In this process, the pressure and temperature is applied simultaneously. Moulding (or compaction) takes place at the same time. It reduces shrinkage and improves properties. This is used for limited applications such as for production of very hard cemented carbide parts. Another type is cold isostatic pressing in which only pressure is applied uniformly to give the desired size and shape to the powder.

4.9 SUPPLEMENTARY OPERATIONS

- This is used, where higher density or close dimensional tolerances are required. The sintered compact may show a slight size difference from designed one. This correction is done by placing the component in a master die and applying pressure. These operations are called as sizing or coining or repressing.

- Presintering involves interruption of sintering process at some temperature. This improves machinability. In some cases, resintering is carried out after re-pressing. This improves mechanical properties due to reduction in shrinkage.

- The post sintering heat treatment as stress relief or annealing may be used in some cases. Age hardening for non-ferrous and surface hardening for ferrous metals may be used. Various finishing operations as machining, shearing, broaching, deburring, grinding etc. are used.

- Impregnation is used to fill internal pores in sintered compact. This is carried out to improve antifriction properties. Various oils, waxes and grease may be used for impregnation. Impregnation with liquid lead is used to improve the specific gravity of ferrous metal compacts.

Testing of P/M Parts

The component should be tested for various properties as compressive and tensile strength, porosity, density, hardness, chemical composition and microstructure etc.

4.10 ADVANTAGES AND LIMITATIONS OF POWDER METALLURGY PROCESSES

4.10.1 Advantages

Powder metallurgy process has several advantages over other conventional shaping processes.

These are summarized as below :

- Without referring to equilibrium or phase diagram, components of any required compositions can be produced.

- A combination of metal powder and non-metallic material is possible.

- A close control on the amount of porosity is possible (See Fig. 4.10). This is required in production of filters, bearing and bricks etc.

- The production of refractory metals like tungsten, tantalum etc. and heavy metals is possible without melting.

- The production of components from metals, which are insoluble in each other during melting is possible by this process.

- Precise control on materials and properties is possible from starting step of production of powder to finishing step.

- From a single material, a range of densities can be produced simply by controlling pressure and temperature.

- High density parts can be produced.

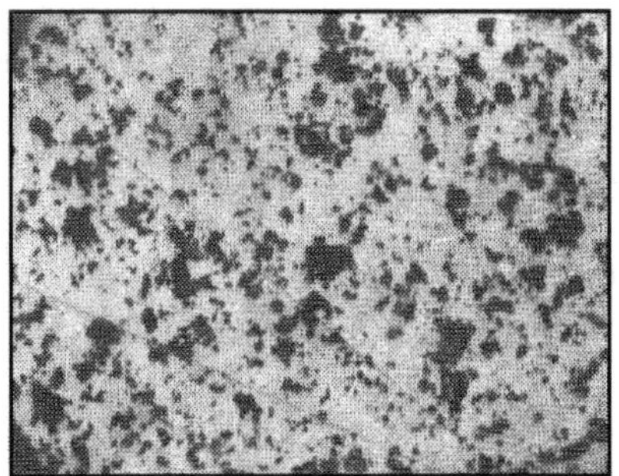

(400 X)

Fig. 4.10: Photomicrograph showing porosity (as black areas) in powder metallurgy component

- More complicated shaped parts as pinion can be satisfactorily made by this process.

- From single piece P/M component space saving and other design advantages can be obtained.

- It eliminates scrap.

- The powder metallurgical parts show good damping characteristics. This is used in air conditioning blowers.

- Practically, any desired material can be mixed.

- Composite and dispersion hardened materials can be produced.

- Production of cemented carbide tool is possible only by this process.

- Purity of metal powder can be retained upto final operations.

- It is an economical process for mass production. It is a very fast and smooth process.

- It gives excellent reproducibility.

- It can offer very complex shapes.

- It eliminates numerous machining operations.

- P/M parts can be welded, soldered or brazed easily, similar to conventional metal parts.

- Highly qualified or skilled worker is not required.

4.10.2 Limitations

P/M process show limitations as explained below :

- The metal powders used in these processes are very fine. These fine powders tend to explode and cause fire hazards and oxidation during storing.

- It is very difficult to produce high purity powder. It is also expensive to maintain purity.

- Alloy powders such as stainless steel, brass, bronze are difficult to produce as simple method is not available.

- Very large sized components cannot be produced because of limitation of capacity of processes and furnaces available.

- For complex parts, the designing and machining of tooling is not easy.

- Components of theoretical density cannot be produced.

- Due to porosity, the specified mechanical properties are difficult to be obtained.

- Porous metals tend to oxidize rapidly.

- P/M parts show comparatively poor plastic properties.

- Higher investment is required for heavy presses and tooling (i.e. punches, dies etc.)

4.11 APPLICATIONS

The P/M parts find applications in the following areas

- **Refractory Metals:** Components produced by using refractory metals as tungsten, molybdenum and tantalum are used in electric light bulb, fluorescent lamps, radio valves, mercury arc rectifiers, x-ray tube etc.

- **Refractory Carbides:** These are very hard materials suitable for machining operations.

  - **Machining:** Lathes, drilling, knurling

  - **Instruments:** Gauges, indentors

  - **Wire Drawing:** Dies, blocks, jaws

  - **Deep Drawing:** Dies

  - **Mining:** Stone hammers, chisels, saws and also in chemical, textile and ceramic industries.

  - **Automotive:** A large group of P/M parts are used in automobile industries. The application includes porous bearings, slide bearings, self-lubricated bearings, sintered and oil pump gears etc. Another important application covers the components like clutch plates, discs, electrical contacts, crankshaft drive or camshaft sprocket etc.

  - **Aerospace Applications:** P/M parts are used in space applications in rocket, missiles, satellite etc. Tungsten parts are used in plasma jet engines at about 1850°C. Sintered bronze bearing and other bushing are used in explorer satellites, magnetic materials such as *Alnico* in communication system. Beryllium is used in gyroscope and heat shield for space capsules. Nickel is used in fuel cell for space electronic system. Be, Al, Mg and Zr are used in the form of solid fuels for rockets and missiles.

  - **Atomic Energy Applications:** Various P/M parts are used in nuclear reactor. Dispersion strengthened materials are used in atomic reactors and rockets, magneto-hydrodynamic generators, high temperature gas turbine etc. Beryllium is used as a fuel canning material for nuclear reactor. Uranium carbide, uranium dioxide are used as a fuel material, Be as moderator, Zr as cladding material.

  - **Defense Applications:** P/M parts and metal powders are used in rockets, missiles, nose piece fuses, cartridge cases, fragile bullets etc.

  - **Other Applications:** Metal filters are used in various chemical, sugar and pharmaceutical industries. Porous nickel electrodes are used in Ni-Cd batteries. Sintered friction materials are used as brakes in various applications. Electrical contacts are used in relays actuators, timers, switch gears etc. The typewriter contains various cams, gears and pawls made by P/M. Sintered tungsten carbide is used as ball on most of the ball point pens. P/M parts are also used as surgical implants.

4.12 SECONDARY FINISHING OPERATIONS

Secondary Operations on Powder Metal Part :

Machining :

- In general, machining is not necessary for porous parts. But it can be done to produce specific shape and size in which case a very sharp tooling with a slight rake is employed.

- The machined surface is then treated to remove the cutting fluids. EDM and laser cutting are also performed to obtain specific shape and size.

Joining :

- Joining of porous parts can be done with one another or with solid part mainly in the case of stainless steel porous components.

- TIG, LASER or electron beam welding are recommended for satisfactory joining of porous parts. Soldering and brazing are not used. Epoxy resins are also used for bonding of porous parts.

- Insert moulding, sinter bonding, press fitting are other secondary operations.

Applications of Porous Parts :

Porous Metallic Filters :

- Porous filters remove solid particles from streams of liquid such as oil, gasoline, refrigerants, polymer melts and from air or other gases.

- The important characteristics of filter materials are, adequate mechanical strength, retention of solid particles up to a specified size, fluid permeability, adequate resistance to atmosphere attack.

- Typical pore sizes are 0.2, 0.5, 1, 2, 5, 10, 20, 40, 60 and 100 microns. The most widely used metallic filter materials are porous copper-tin bronze and porous stainless steel.

- For filtering highly corrosive fluids, filters made form monel, inconel, Ti can be used.

- They are also used in flow control devices, distribution applications etc.

4.13 DESIGN CONSIDERATIONS

Economics Usually Require Large Quantities to Justify Cost of Equipment and Special Tooling

- Minimum quantities of 10,000 units are suggested.

- PM is unique in its capability to fabricate parts with a controlled level of porosity.

- Porosities up to 50% are possible.

- PM can be used to make parts out of unusual metals and alloys - materials that would be difficult if not impossible to produce by other means.

The Part Geometry Must Permit Ejection from Die After Pressing

- This generally means that part must have vertical or near-vertical sides, although steps are allowed.

- Design features such as undercuts and holes on the part sides must be avoided.

- Vertical undercuts and holes are permissible because they do not interfere with ejection.

- Vertical holes can be of cross-sectional shapes other than round without significant difficulty.

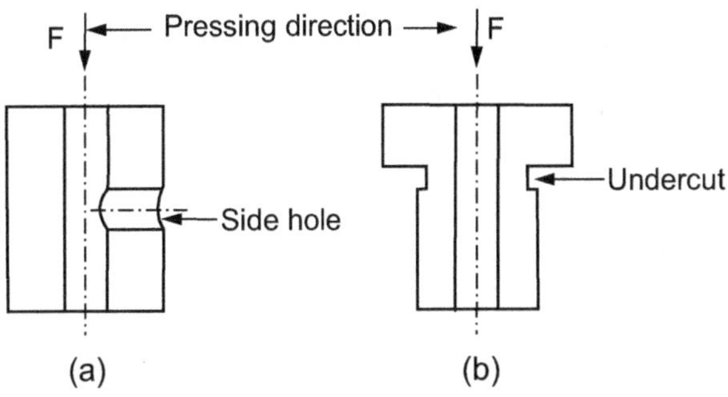

Fig. 4.11 : Part features to be avoided in PM: (a) side holes and (b) side undercuts since part ejection is impossible

- Screw threads cannot be fabricated by PM; if required, they must be machined into the part. Chamfers and corner radii are possible by PM pressing, but problems arise in punch rigidity when angles are too acute.

- Wall thickness should be a minimum of 1.5 mm (0.060 in) between holes or a hole and outside wall.

- Minimum recommended hole diameter is 1.5 mm (0.060 in).

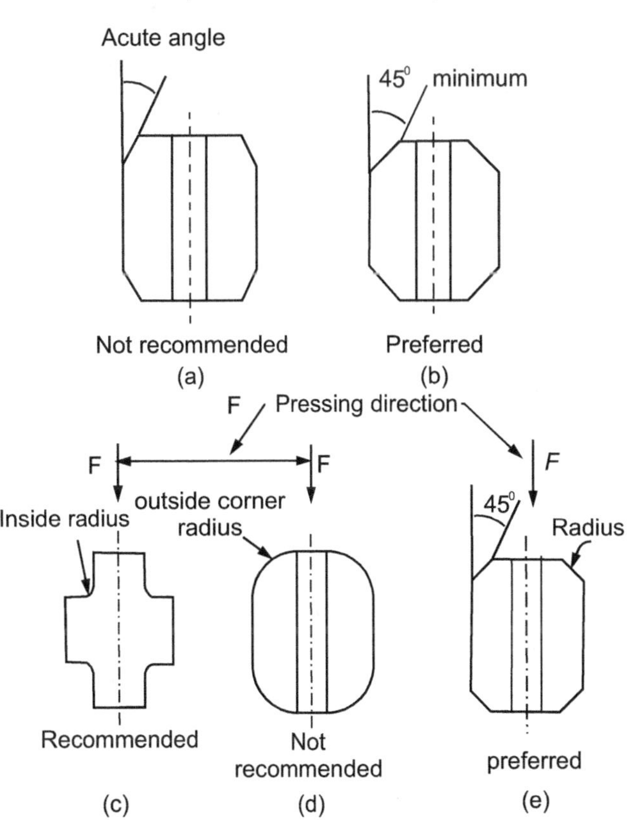

Fig. 4.12 : Chamfers and corner radii are accomplished but certain rules should be observed: (a) avoid acute angles; (b) larger angles preferred for punch rigidity; (c) inside radius is desirable; (d) avoid full outside corner radius because punch is fragile at edge; (e) problem solved by combining radius and chamfer

- Theoretically, sharp corners and edges facing outward can be produced. In practice, however, it would be preferable to round them off the die will then be easier to design and will also be less susceptible to cracking. If the die is designed in one single piece, a minimum radius will always be present. This radius is generated by the tools used to carve the die.

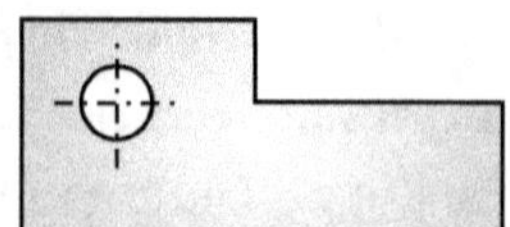

(a) Possible

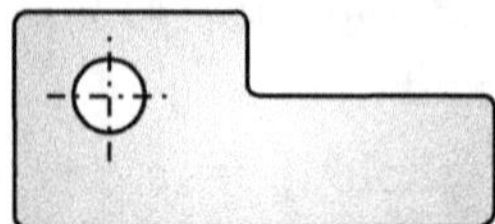

(b) Preferred

Fig. 4.13 : Corners and edges facing the die

EXERCISE

1. Why the factors like particle size, shape and distribution have importance in the packing of powder ?

2. Discuss the effect of powder production method on powder shape.

3. Suggest suitable metals and the production process for the following
   (a) Friction materials,
   (b) Electrical contacts.

4. Explain why P/M process is unsuitable for certain components. Give 3-4 such components and justify.

5. List (do not discuss) the various applications of P/M process.

6. Explain any four powder production processes.

7. Discuss in general the process of powder metallurgy with respect to the following points
   (a) Powder production,
   (b) Compaction and
   (c) Sintering.

9. What are the important characteristics of metal powders ? How are they evaluated ?

10. Discuss the advantages and disadvantages of powder metallurgy.

11. Write note on powder conditioning.

PROCESSING OF CERAMICS AND GLASSES

5.1 INTRODUCTION

- Ceramics form an important part of materials group. Ceramics are compounds between metallic and nonmetallic elements for which the inter-atomic bonds are either ionic or predominantly ionic. The term ceramics comes from the Greek word keramikos which means 'burnt stuff'. Characteristic properties of ceramics are, in fact, optimized through thermal treatments. They exhibit physical properties those are different from that of metallic materials.

- The term "Ceramics Processing" describes the process of production of ceramic components from natural to synthetic raw materials as well as their disposal.

- Contrary to metals, polymers or glasses, the starting materials for the production of ceramic materials are powders. These powders are brought into shape and the components then are sintered which is a temperature treatment clearly below the melting point. This technique is applied because of the high melting points of ceramic materials which make casting impossible or uneconomical. The starting material can be of oxidic or non-oxidic nature; some care must be taken in order not to sinter both types of materials in the same furnace. Example: silicon nitride would incinerate or combust if sintered in an oxidizing atmosphere. Therefore, furnace technology for non-oxidic materials must be different from the one for oxidic starting materials.

- Classification of the most important material groups distinguishes between natural and synthetic materials. Natural raw materials are extracted from earth, and these raw materials must be further processed. They are blast in quarries, i.e., pieces of rock are exploited and worked up to powders (materials). From these materials pre-products are produced by forming or shaping. Metal components can be formed during the process of reshaping whereas ceramic components can be produced only by a sintering process. The product in later time has to be disposed by recycling or remineralisation.

5.2 TYPES AND APPLICATIONS OF CERAMICS

- Classification of ceramics based on their specific applications and composition are two most important ways among many. Based on their composition, ceramics are classified as:

  ➢ Oxides,

  ➢ Carbides,

  ➢ Nitrides,

  ➢ Sulfides,

  ➢ Fluorides, etc.

The other important classification of ceramics is based on their application, such as:

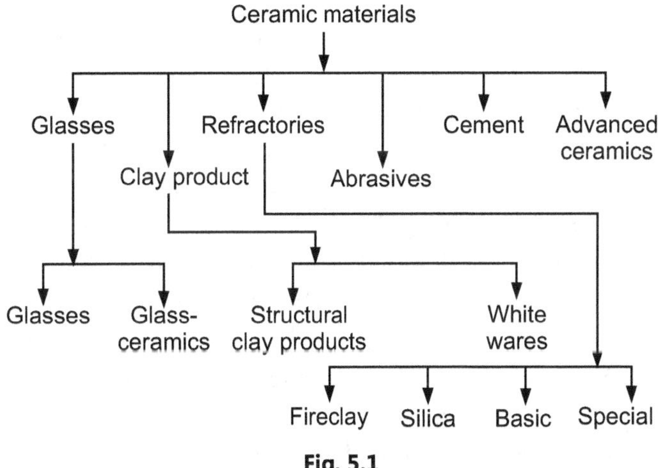

Fig. 5.1

- In general, ceramic materials used for engineering applications can be divided into two groups: traditional ceramics, and the engineering ceramics. Typically, traditional ceramics are made from three basic components: clay, silica (flint) and feldspar. For example bricks, tiles and porcelain articles. However, engineering ceramics consist of highly pure compounds of aluminium oxide (Al_2O_3), silicon carbide (SiC) and silicon nitride (Si_3N_4).

- **Glasses:** Glasses are a familiar group of ceramics – containers, windows, mirrors, lenses, etc. They are non-crystalline silicates containing other oxides, usually CaO, Na_2O, K_2O and Al_2O_3 which influence the glass properties and its color. Typical property of glasses that is important in engineering applications is its response to heating. There is no definite temperature at which the liquid transforms to a solid as with crystalline materials. A specific temperature, known as glass transition temperature or fictive temperature is defined based on viscosity above which material is named as super cooled liquid or liquid, and below it is termed as glass.

- **Clay Products:** Clay is the one of most widely used ceramic raw material. It is found in great abundance and popular because of ease with which products are made. Clay products are mainly two kinds – structural products (bricks, tiles, sewer pipes) and white-wares (porcelain, chinaware, pottery, etc.).

- **Refractories:** These are described by their capacity to withstand high temperatures without melting or decomposing; and their inertness in severe environments. Thermal insulation is also an important functionality of refractories.

- **Abrasive Ceramics:** These are used to grind, wear, or cut away other material. Thus the prime requisite for this group of materials is hardness or wear resistance in addition to high toughness. As they may also exposed to high temperatures, they need to exhibit some refractoriness. Diamond, silicon carbide, tungsten carbide, silica sand, aluminium oxide / corundum are some typical examples of abrasive ceramic materials.

- **Cements:** Cement, plaster of paris and lime come under this group of ceramics. The characteristic property of these materials is that when they are mixed with water, they form slurry which sets subsequently and hardens finally. Thus it is possible to form virtually any shape. They are also used as bonding phase, for example between construction bricks.

- **Advanced Ceramics:** These are newly developed and manufactured in limited range for specific applications. Usually their electrical, magnetic and optical properties and combination of properties are exploited. Typical applications: heat engines, ceramic armors, electronic packaging, etc.

5.3 PROCESSING OF CERAMICS

- Ceramics melt at high temperatures and they exhibit a brittle behavior under tension. As a result, the conventional melting, casting and thermo-mechanical processing routes are not suitable to process the polycrystalline ceramics.

- Inorganic glasses, though, make use of lower melting temperatures due to formation of eutectics. Hence,

most ceramic products are made from ceramic powders through powder processing starting with ceramic powders.

- The powder processing of ceramics is very close to that of metals, powder metallurgy. However there is an important consideration in ceramic-forming that is more prominent than in metal forming: it is dimensional tolerance.

- Post forming shrinkage is much higher in ceramics processing because of the large differential between the final density and the as-formed density.

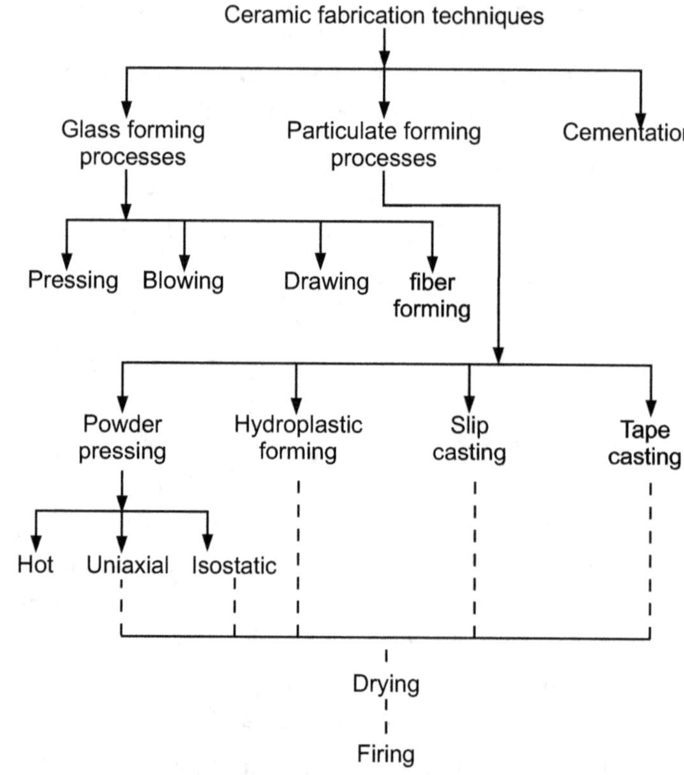

Fig. 5.2

5.4 SHAPING CERAMICS

- Ceramic powder processing consists of powder production by milling/grinding, followed by fabrication of green product, which is then consolidated to obtain the final product.

- A powder is a collection of fine particles. Synthesis of powder involves getting it ready for shaping by crushing, grinding, separating impurities, blending different powders, drying to form soft agglomerates. Different techniques such as compaction, tape casting, slip casting, injection moulding and extrusion are then used to convert processed powders into a desired shape to form what is known as green ceramic.

- The green ceramic is then consolidated further using a high-temperature treatment known as sintering or firing.

5.4.1 Slip Casting

A slip is a suspension of clay and / or other nonplastic materials in water.

- Preparation of a powdered ceramic material and a liquid (usually clay and water) into a stable suspension called a slip.
- Pouring the slip into a porous mould usually made of plaster of paris and allowing the liquid portion of the slip to be partially absorbed by the mould. As the liquid is removed from the slip, a solid layer is formed against the mould surface.
- When the sufficient wall thickness has been formed, the casting process is interrupted and the excess slip is poured out of the cavity (Drain casting).
- The material in the mould is allowed to dry to provide adequate strength for handling and the subsequent removal of the part from the mould.

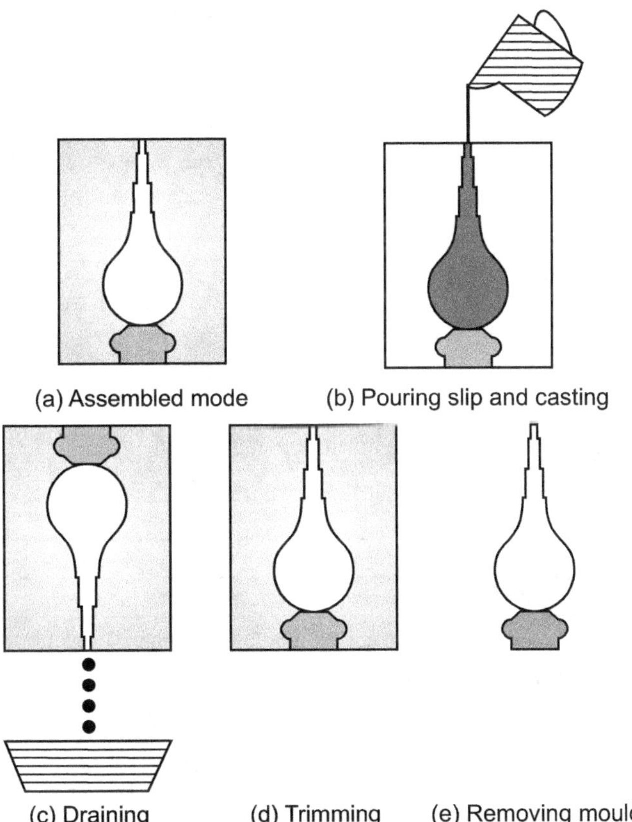

Fig. 5.3 : Slip casting steps

- This process may be continued until the entire mould cavity becomes solid (solid casting), as Or it may be terminated when the solid shell wall reaches the desired thickness, by inverting the mould and pouring out the excess slip; this is termed drain casting. As the cast piece dries and shrinks, it will pull away (or release) from the mould wall; at this time the mould may be disassembled and the cast piece removed.

- The nature of the slip is extremely important; it must have a high specific gravity and yet be very fluid and pourable. These characteristics depend on the solid to-water ratio and other agents that are added. A satisfactory casting rate is an essential requirement. In addition, the cast piece must be free of bubbles, and it must have a low drying shrinkage and a relatively high strength.

- The properties of the mould itself influence the quality of the casting. Normally, plaster of paris, which is economical, relatively easy to fabricate into intricate shapes, and reusable, is used as the mould material. Most moulds are multipiece items that must be assembled before casting.

- Also, the mould porosity may be varied to control the casting rate.The rather complex ceramic shapes that may be produced by means of slip casting include sanitary lavatory ware, art objects, and specialized scientific laboratory ware such as ceramic tubes.

Advantages of Slip Casting:

- Produced uniform thickness for complex shapes
- Economic to develop parts with short production runs.

5.4.2 Tape Casting

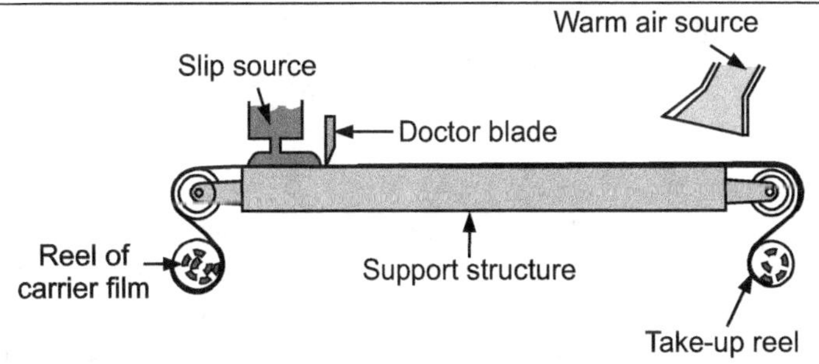

Fig. 5.4 : Tape casting

- An important ceramic fabrication technique, tape casting. As the name implies, thin sheets of a flexible tape are produced by means of a casting process. These sheets are prepared from slips, in many respects similar to those that are employed for slip casting. This type of slip consists of a suspension of ceramic particles in an organic liquid that also contains binders and plasticizers that are incorporated to impart strength and flexibility to the cast tape. De-airing in a vacuum may also be necessary to remove any entrapped air or solvent vapor bubbles, which may act as crack-initiation sites in the finished piece.

- The actual tape is formed by pouring the slip onto a flat surface (of stainless steel, glass, a polymeric film, or paper); a doctor blade spreads the slip into a thin tape of uniform thickness, as shown schematically in Fig. 5.4

- In the drying process, volatile slip components are removed by evaporation; this green product is a flexible tape that may be cut or into which holes may be punched prior to a firing operation. Tape thicknesses normally range between 0.1 and 2 mm (0.004 to 0.08 in.). Tape casting is widely used in the production of ceramic substrates that are used for integrated circuits and for multilayered capacitors.

5.4.3 Pressure Casting

- Pressure casting has first been developed for manufacturing of voluminous thick-walled components in sanitary ceramics like for example sinks, but later-on this process has been transferred to other ceramic materials and components, as well. Applying pressure during the casting process will reduce the casting time required and allow for different wall thicknesses in the solid casting process.

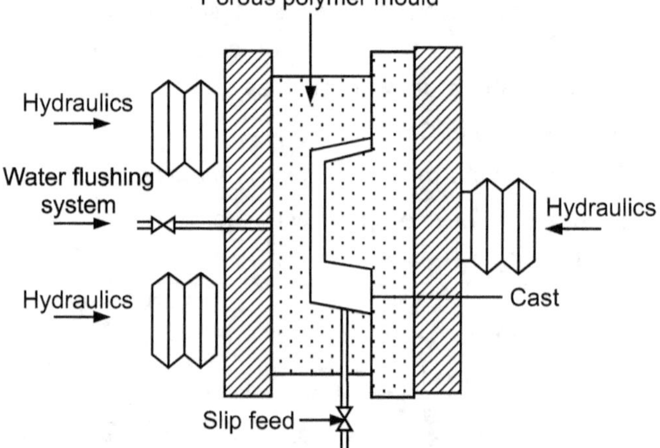

Fig. 5.5 : Pressure casting (schematic)

- Porous normally non sucking polymers are used as materials for the moulds. Upon formation of the ceramic cast the porous polymer mould is opened and is rinsed with water from behind to release the cast. As the mould material must not be self-drawing it needs not to be dried. For pressures up to about 40 bars that are usually applied in pressure casting.
- Because of the varying compressibility of ceramic materials the pressure dependence of the casting rate must be determined for each system separately. For a ZrO_2 slip with 83 wt.% solid yield the pressure dependence of the cast thickness.
- An up-to-date casting plant several polymer moulds are connected to each other so that up to ten moulds may be filled simultaneously according to the component geometry required. The polymer moulds allow for up to 20.000 casts subject to the slip quality used. Besides closer dimensional tolerances and reduced casting times at variable and higher wall thicknesses the essentially increased durability of the moulds offers considerable advantages compared to the conventional slip casting process at atmospheric conditions.
- The decision for the respective casting process is ultimately subject to the quantities to be produced and the economic contemplations involved. Further pressure-supported casting processes that are applied for special component and material groups are the pressure filtration (hard magnetic ferrites), the vacuum casting (fibre insulation) and the centrifugal casting (high tech ceramics).

5.4.4 Plastic Forming

- During the forming, deflocculated slurries, plasticized materials or granules are transformed into green bodies with defined size, shape, density and reproducible tolerances. The geometric dimensions required and the quantities to be produced are decisive for determining the forming process to be applied. The shrinkage caused by the subsequent firing process is subject to the fluctuations of the green density and the dimensional tolerances.
- The reproducibility of these sizes and the avoidance of defects that can hardly be cured any longer in the subsequent sintering decide on the economy of the respective forming process apart from investment and personnel costs. In order to achieve a sufficient green strength for the transport of the parts, but also to optimize the processability characteristics of the initial materials for the respective forming process, organic additives are added to the ceramic feeds to fulfil different functions

Functions of processing additives in ceramic.

Additive	Function
Ceramic powder	Matrix
Sintering additive	Densification aid
Solvent	Dispersion
Deflocculant	Control of surface charges and pH, dispersion
Dispersing agent	Deagglomeration
Wetting agent	Reduce of the surface tension of the solvent
Antifoaming agent	Avoidance of bubbles
Preservative	Avoidance of bacteria cultures
Binder	Green strength
Plasticizer	Flexibility
Softener	Flexibility
Lubricant	Reduce die and internal friction, mould release

5.4.5 Extrusion

- Extrusion of plastic ceramic feeds is used for manufacturing components with defined cross sections whose length is determined by cutting an extruded rod. The plastic material is fed through a charging hopper and the metering screw into the press. In a vacuum chamber the feed is evacuated and fed by the auger to the die (Fig. 5.6).

- Depending on the die geometry, feeding extrudates are produced to be further process as starting material for the manufacturing of isolators and dinner ware or to extrude bricks, tiles, tubes, substrates or "honeycomb" structures for catalyst carriers by means of suitable inserts.

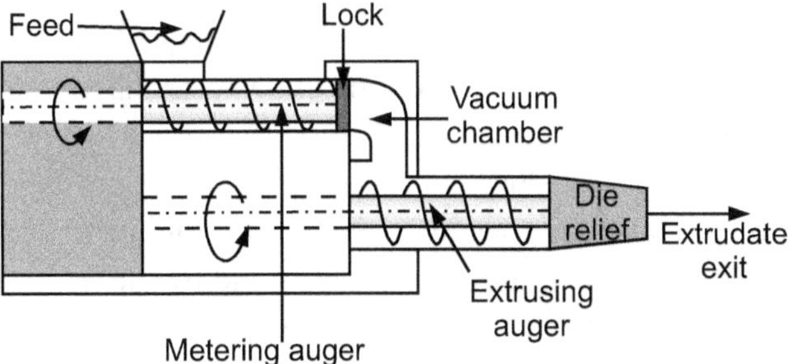

Fig. 5.6 : Extrusion forming

- In case of clay based feeds the raw materials are plasticized by water and very small quantities of defloculants. Oxide and non-oxide ceramic raw materials are plasticized by aqueous and non-aqueous binder systems.
- For the practical use the mass flow and the pressure course around the die of the extruder are important. At the transition between the auger cylinder and the die, a pressure drop is caused that reflects the energy required to deform the material towards the smaller diameter. The pressure drop in the die reflects the stress required to overcome the friction forces so that the material glides. The pressure drop depends on the rheological behaviour of the feed and on the geometric conditions of the extruder.
- The complex pressure conditions and relative motions inside the extruder cause textures in the extruded parts. Especially in clay based feeds micro-textures are built up due to the particle alignment.

5.4.6 Injection Moulding

- Injection moulding is a process to manufacture small components of complex geometries and low wall thicknesses in large quantities. Typical injection moulded components are cores for metal casting, thread guides, cutting tools, welding nozzles and turbocharger rotors.
- Ceramic powders with binders, plasticizers and lubricants are homogenized to plastify the feeds, which is done in heatable mixers or kneaders above the melting point of the organic additives. The high shear stresses required for mixing and for the dispersion of agglomerates will result in high torques in the mixer

and often in considerable abrasion of metallic components.

- The homogenized feed with up to 50 vol.% of organic additives is cooled and granulated simultaneously through the screws or sigma rotors. This granulate is fed through the filling hopper to the heatable injection moulding machine.

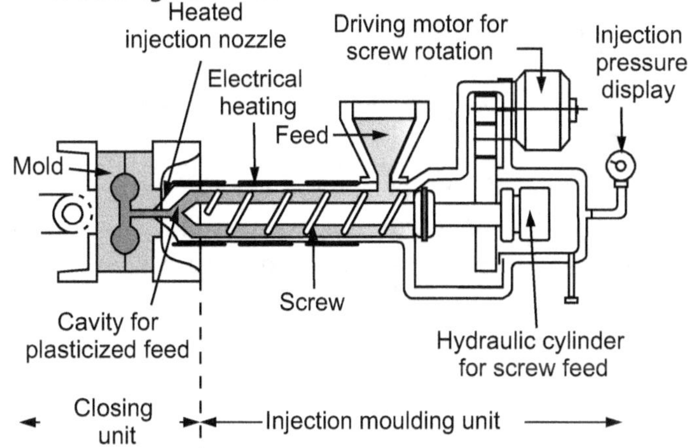

Fig. 5.7 : Schematic diagram of a screw type injection moulding machine

- The feed is plasticized in the heated cylinder and injected into the mould by the screw which acts both as feeding and pressing device. The temperature zones of the heatable plasticizing cylinder are subject to the melting points and the viscosity course of the organic constituents used. They range between 440 and 510 K depending on the flow behaviour of the powders and the content of organic components.
- The mould temperature is lower than the melting points of the organic materials used so that the plastic feed solidifies and may be removed from the mould. It should be adjusted so as to maintain the flowability of the feed till the mould is completely filled. High mould temperatures may thus avoid press cones and inhomogenities in the sample cross section. On the other hand this temperature must be far enough below the melting point of the lowest-melting organic constituent so that the formation of bubbles at the sample surfaces will be avoided upon removal from the mould.
- According to the flow behaviour of the feed screws with different geometries are used. For a high organic content and powders with a wide grain size distribution conventional screws for thermoplastics with compression zone and sprue may be used because of the good flowability of the mixtures. In the compression zone in the front part of the screw the

core diameter increases so that an additional homogenizing effect is obtained. The sprue prevents back flowing material during the injection process.

- When using powders with a very narrow grain size distribution, or in case of very low organic contents in the feed, duromer screws without compression zone and sprue are used because of the bad flowability of the mixtures. The homogenity of the injection moulded samples may in addition be improved by optimizing the injection velocity and the injection pressure.

5.4.7 Pressing

- The term of dry pressing generally describes the densification of powders or granules in axial direction between two stamps in hardened metal moulds.

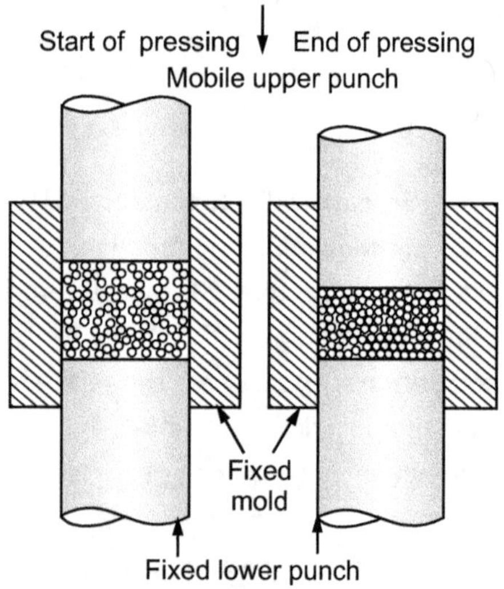

Fig. 5.8 : Schematic diagram of the uniaxial dry pressing process

- This process allows only comparatively simple geometries to be manufactured like magnetic and dielectric components, alumina substrates, cutting tools, sensors, refractories, grinding disks, steatite isolators and tiles. After filling the mould the parts are densified and ejected. The densification is generally described by a pressure-density curve and is divided into three ranges. In the first range of densification the particles are displaced in the pressing direction and densified by sliding and orienting. The sliding processes are favoured by organic additives that are normally added to the powders before or during the granulation.

- The total concentration of densification aids ranges between 1 and 5wt.%, as a rule. In the second range agglomerates are destroyed, in the third plastic deformation is partly obtained. The density after pressing increases in the first range almost linear with the operating pressure, but shows a highly progression after all.

- For Al_2O_3 fluid bed sprayed granules with polyvinyl alcohol (PVA) and polyethyleneglycol (PEG) as lubricant they showed that the compressibility curve may be pre-calculated very well from the characteristics of the individual granules assuming certain simplifications like for example incompressibility of the individual granules.

- All these contemplations start from a uniform densification of the component cross section. Subject to the diameter/height ratio, however, the obtained density distributions cause a deformation of the components during sintering. This non-uniform densification is caused by friction between the grains, friction between powder particles and mould as well as by the different displacements of the individual granules. In practical use these factors are opposed by changing the organic additives and by tuning the movement of the upper and the lower punch.

5.5 DRYING AND FIRING

- A ceramic piece that has been formed hydroplastically or by slip casting retains significant porosity and insufficient strength for most practical applications. In addition, it may still contain some liquid (e.g., water), which was added to assist in the forming operation. This liquid is removed in a drying process; density and strength are enhanced as a result of a high-temperature heat treatment or firing procedure.

- A body that has been formed and dried but not fired is termed green. Drying and firing techniques are critical in as much as defects that ordinarily render the ware useless (e.g., warpage, distortion, and cracks) may be introduced during the operation. These defects normally result from stresses that are set up from nonuniform shrinkage.

5.5.1 Drying

- As a clay-based ceramic body dries, it also experiences some shrinkage. In the early stages of drying the clay particles are virtually surrounded by and separated from one another by a thin film of water. As drying progresses and water is removed, the interparticle separation decreases, which is manifested as shrinkage. During drying it is critical to control the rate of water removal.

- Drying at interior regions of a body is accomplished by the diffusion of water molecules to the surface where evaporation occurs. If the rate of evaporation is greater than the rate of diffusion, the surface will dry (and as a consequence shrink) more rapidly than the interior, with a high probability of the formation of the aforementioned defects. The rate of surface evaporation should be diminished to, at most, the rate of water diffusion; evaporation rate may be controlled by temperature, humidity, and the rate of airflow.

- Other factors also influence shrinkage. One of these is body thickness; nonuniform shrinkage and defect formation are more pronounced in thick pieces than in thin ones. Water content of the formed body is also critical: the greater the water content, the more extensive the shrinkage. Consequently, the water content is ordinarily kept as low as possible. Clay particle size also has an influence; shrinkage is enhanced as the particle size is decreased.

- To minimize shrinkage, the size of the particles may be increased, or nonplastic materials having relatively large particles may be added to the clay. Microwave energy may also be used to dry ceramic wares.

- One advantage of this technique is that the high temperatures used in conventional methods are avoided; drying temperatures may be kept to below this is important because the drying of some temperature-sensitive materials should be kept as low as possible.

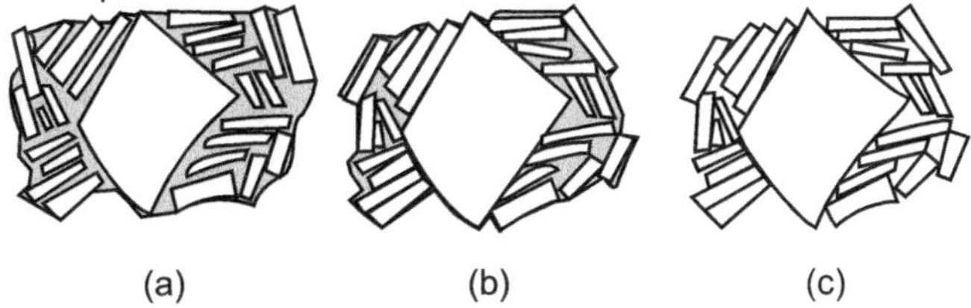

(a)   (b)   (c)

Fig. 5.9 : Several stages in the removal of water from between clay particles during the drying process. (a) Wet body. (b) Partially dry body. (c) Completely dry body

5.5.2 Firing

- After drying, a body is usually fired at a temperature between 900°C to 1400°C the firing temperature depends on the composition and desired properties of the finished piece. During the firing operation, the density is further increased (with an attendant decrease in porosity) and the mechanical strength is enhanced. When clay-based materials are heated to elevated temperatures, some rather complex and involved reactions occur. One of these is vitrification, the gradual formation of a liquid glass that flows into and fills some of the pore volume. The degree of vitrification depends on firing temperature and time, as well as the composition of the body.

- The temperature at which the liquid phase forms is lowered by the addition of fluxing agents such as feldspar. This fused phase flows around the remaining unmelted particles and fills in the pores as a result of surface tension forces (or capillary action); shrinkage also accompanies this process. Upon cooling, this fused phase forms a glassy matrix that results in a dense, strong body. Thus, the final microstructure consists of the vitrified phase, any unreacted quartz particles, and some porosity.

- The degree of vitrification, of course, controls the room-temperature properties of the ceramic ware; strength, durability, and density are all enhanced as it increases. The firing temperature determines the extent to which vitrification occurs; that is, vitrification increases as the firing temperature is raised. Building bricks are ordinarily fired around 900°c and are relatively porous. On the other hand, firing of highly vitrified porcelain, which borders on being optically translucent, takes place at much higher temperatures. Complete vitrification is avoided during firing, since a body becomes too soft and will collapse.

5.6 FINISHING OPERATIONS

Machining (Grinding, Drilling, Sawing)

- Machining operations may be necessary to produce ceramic products whose final shape or dimensional tolerance cannot be achieved technically or with sufficient accuracy during primary processing (especially for larger shapes or blocks).

Wet Grinding

- Wet grinding is utilised to finish products requiring the tightest dimensional tolerance. Grinding is a batch process, in which a number of pieces are fixed to a table, which traverses under a diamond machining head. Bed surfaces of building bricks or blocks are sometimes smoothed by wet grinding to facilitate bonding with thin-layer 'glues'.

Dry Grinding

- Dry grinding of clay block bed surfaces with diamond wheel grinding systems is carried out as subsequent treatment to facilitate bonding with a thin layer of mortar. In the case of dry grinding, the whole grinding device is encapsulated.

Drilling

- Ceramic products, especially refractory products may need to be drilled when the 'hole' required cannot be achieved with the necessary accuracy during the pressing and firing operations.

Sawing

- This is a finishing operation used when the final shape of a ceramic brick, especially a refractory brick, cannot be effectively produced at the pressing stage. In this situation, oversize bricks are pressed and fired, then sawn to the required dimensions. Facing bricks may be sawn in the manufacture of 'cut and stuck' special shapes. In virtually all machining operations, a closed loop water system provides lubrication and sweeps lubricated particles away from the working surfaces – at the same time, minimising dust release.

Polishing

- In some cases, particularly involving porcelain tiles, the fired surface is polished to achieve a shiny, unglazed homogenous tile.

Carbon Enrichment (Refractory Products)

- Refractory products are required to work in extremely hostile working environments, and for certain applications it is necessary to impregnate fired ware with petroleum-based pitch. The presence of carbon in the final product offers several advantages:

  ➢ It acts as a lubricant, beneficial for the working surfaces of sliding-gate plates

  ➢ the relatively high thermal conductivity of carbon increases the thermal shock resistance of the product

  ➢ the carbon acts as a pore filter reducing the permeability of the product, which in turn offers increased resistance to penetration by slag and metal.

- Pitch impregnation is a batch process, typically carried out in three upright cylindrical vessels fitted with hinged lids. Products for treatment are loaded into metal baskets, which fit inside the vessels. The ware is heated in the first vessel to ~ 200 °C via a circulating stream of hot air, then the basket and contents transferred to the second vessel (referred to as an autoclave), which has a heating jacket to maintain the temperature. The autoclave is then sealed, evacuated and filled with liquid pitch (drawn from bulk storage tanks maintained at temperatures of around 180 to 200 °C). Impregnation is achieved by releasing the vacuum, then applying nitrogen at an elevated pressure. After draining, the basket and contents are transferred to the third vessel for cooling to a temperature below that at which volatile pitch components may be evolved.

- Finally, it is necessary to remove a high proportion of pitch volatiles, which would adversely affect the working environment when the refractory products were put into service. Typically, this is achieved by transferring the impregnated ware to an oven, to undergo a defined heating cycle. The oven exhaust fan discharges into a thermal oxidiser (incinerator) held above 800 °C, with a dwell-time of at least 0.5 seconds. These conditions ensure that all pitch volatiles (Complex hydrocarbons) are fully oxidised. Products

which have been through the above treatment emerge coated in light, brittle carbonaceous deposits which must be removed prior to packaging or further processing. This is usually achieved by short blasting the products on a blasting table.

5.7 FORMING AND SHAPING OF GLASS

- Glass is processed by melting and then shaping it, either in moulds, with tools, or by blowing. Glass shapes produced include flat sheets and plates, rods, tubing, glass fibers, and discrete products such as bottles, light bulbs, and headlights. Glass products may be as thick as those for large telescope mirrors. The strength of glass can be improved by thermal and chemical treatments or by laminating it with a thin sheet of tough plastic.

- Glass products generally can be categorized as follows:

  ➢ Flat sheets or plates ranging in thickness from about 0.8 to 10 mm (0.03 to 0.4 in.), such as window glass, glass doors, and tabletops.

  ➢ Rods and tubing used for chemicals, neon lights, and decorative artifacts.

  ➢ Discrete products such as bottles, vases, headlights, and television tubes.

  ➢ Glass fibers to reinforce composite materials and for use in fiber optics.

5.7.1 Flat-Sheet and Plate Glass

- Flat-sheet glass can be made by the float glass method or by drawing or rolling it from the molten state (all three methods are continuous processes):

  ➢ In the **Float Method**, molten glass from the furnace is fed into a long bath in which the glass under a controlled atmosphere and at a temperature of 1150°C (2100°F)—floats over a bath of molten tin. The glass then moves at a temperature of about 650°C (1200°F) over rollers into another chamber, where it solidifies. Float glass has smooth (fire-polished) surfaces, so further grinding or polishing is not necessary. The width can be as much as 4 m. Both thin and plate glass are made by this process.

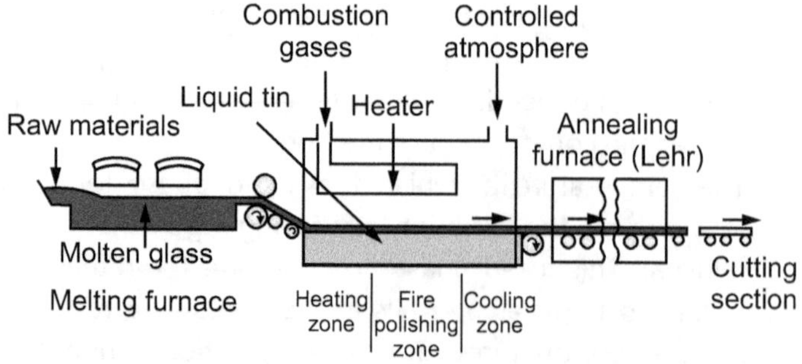

Fig. 5.10 : The float method of forming sheet glass

➢ The **drawing** process for making flat sheets or plates involves passing the molten glass through a pair of rolls in an arrangement similar to an old-fashioned clothes wringer. The solidifying glass is squeezed between these two rolls (forming it into a sheet) and then moved forward over a set of smaller rolls.

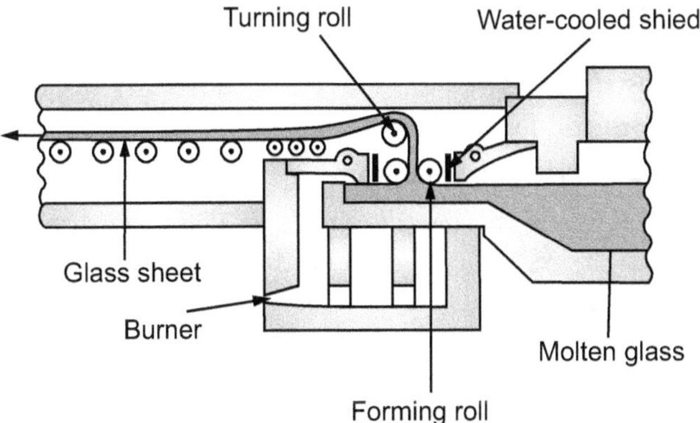

Fig. 5.11 : Drawing process for drawing sheet glass from a molten bath

➢ In the **rolling** process the molten glass is squeezed between powered rollers, thereby forming a sheet. The surfaces of the glass may be embossed with a pattern by using textured roller surfaces. In this way, the glass surface becomes a replica of the roll surface. Thus, glass sheet produced by drawing or rolling has a rough surface appearance.

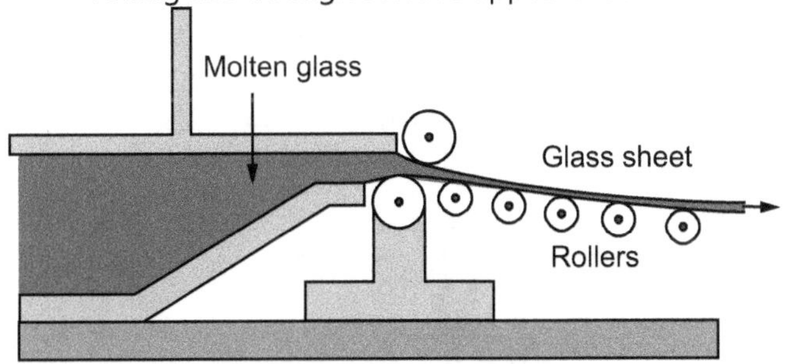

Fig. 5.12 : Rolling process

5.7.2 Tubing and Rods

• Glass tubing is manufactured by the process shown in Fig. 5.13 Molten glass is wrapped around a rotating (cylindrical or cone-shaped) hollow mandrel and is drawn out by a set of rolls. Air is blown through the mandrel to prevent the glass tube from collapsing. These machines may be horizontal, vertical, or slanted downward.

• This is the method used in making the glass tubes for fluorescent bulbs, with machines (such as the Corning Ribbon Machine) capable of producing bulbs at rates of 2000 per minute.

• Glass rods are made in a similar manner, but air is not blown through the mandrel. The drawn product becomes a solid glass rod. An alternative process for making tubes involves extrusion of a strip of glass, which is then wrapped obliquely around a rotating mandrel. The molten glass blends across adjacent layers, and the resultant tube is drawn off the mandrel in a continuous process.

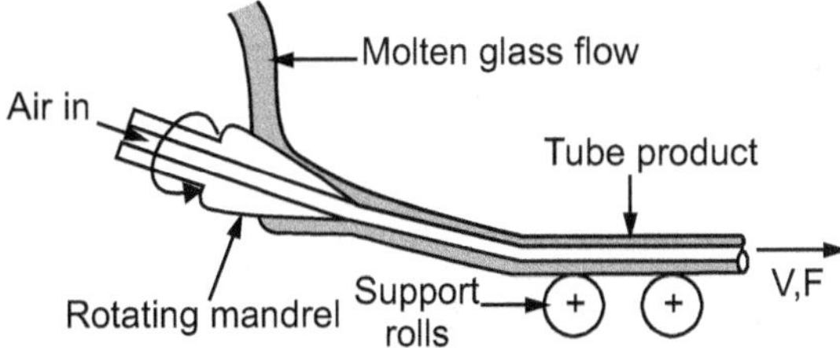

Fig. 5.13 : Manufacturing process for glass tubing

• Air is blown through the mandrel to keep the tube from collapsing. Glass tubes for fluorescent bulbs are made by this method.

5.8 DISCRETE GLASS PRODUCTS

• Several processes are used to make discrete glass objects.

5.8.1 Blowing

➢ Hollow and thin-walled glass items (such as bottles, vases) are made by blowing a process that is similar to the blow moulding of thermoplastics. The steps involved in the production of an ordinary glass bottle by the blowing process are shown in Fig. 5.14.

➢ Blown air expands a hollow gob of heated glass against the inner walls of the mould. The mould usually is coated with a parting agent (such as oil or emulsion) to prevent the glass from sticking to the mould.

➢ Blowing may be followed by a second blowing operation for finalizing product shape, called the blow and blow process.

➢ The surface finish of products made by the blowing process is acceptable for most applications, such as bottles and jars.

➢ It is difficult to control the wall thickness of the product, but this process is economical for high-rate production. Incandescent light bulbs are made in highly automated blowing machines at a rate of greater than 2000 bulbs per minute.

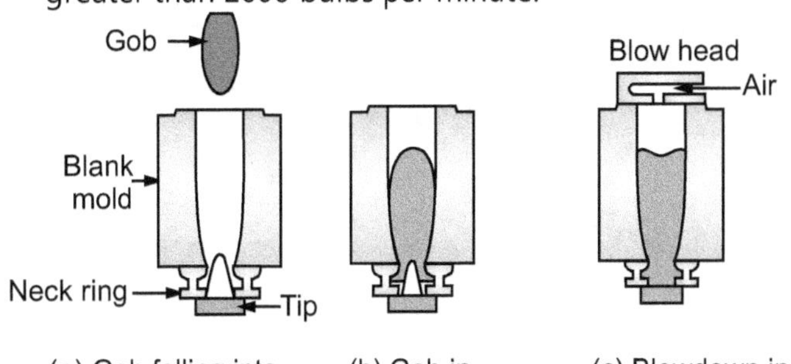

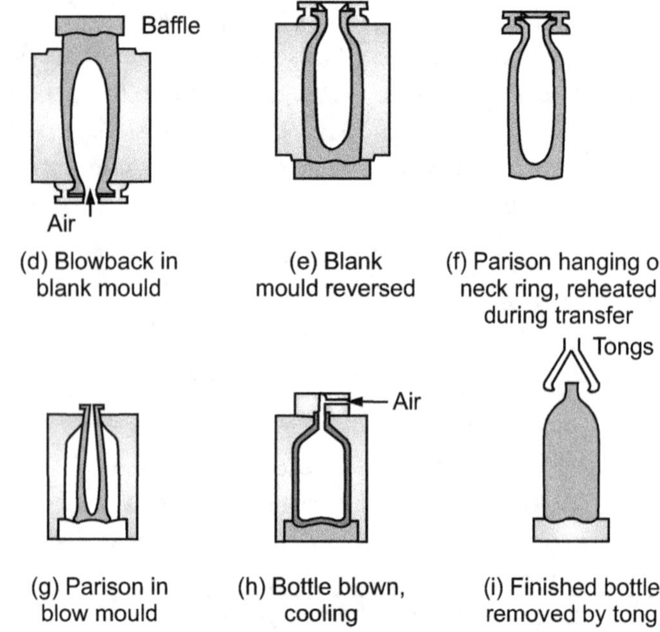

Fig. 5.14 : Steps in manufacturing glass bottle

5.8.2 Pressing

- In the pressing process, a gob of molten glass is placed into a mould and pressed into a confined cavity with a plunger.

- The mould may be made in one piece, or it may be a split mould.

- After being pressed, the solidifying glass acquires the shape of the mould-plunger cavity.

- Because of the confined environment, the product has a higher dimensional accuracy than can be obtained with blowing.

- Split moulds are used for bottles, while, for thin-walled items, pressing can be combined with blowing.

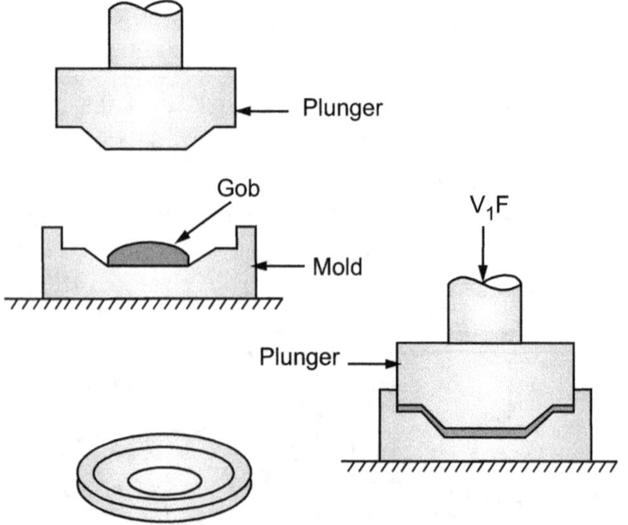

Fig. 5.15 : Manufacturing a glass item
by pressing glass into a mould

5.9 FORMING OF GLASS FIBERS

Glass fiber products can be divided into two categories, with different production methods for each:

1. Fibrous glass for thermal insulation, acoustical insulation, and air filtration, in which the fibers are in a random, wool-like condition.
   Produced by Centrifugal Spraying

2. Long continuous filaments suitable for fiber reinforced plastics, yarns, fabrics, and fiber optics
   Produced by Drawing

5.9.1 Centrifugal Spraying

- In a typical process for making glass wool, molten glass flows into a rotating bowl with many small orifices around its periphery

- Centrifugal force causes the glass to flow through the holes to become a fibrous mass suitable for thermal and acoustical insulation

5.9.2 Drawing of Glass

- Continuous glass fibers of small diameter (lower limit is about 0.0025 mm) are produced by drawing (pulling) strands of molten glass through small orifices in a heated plate made of a platinum alloy.

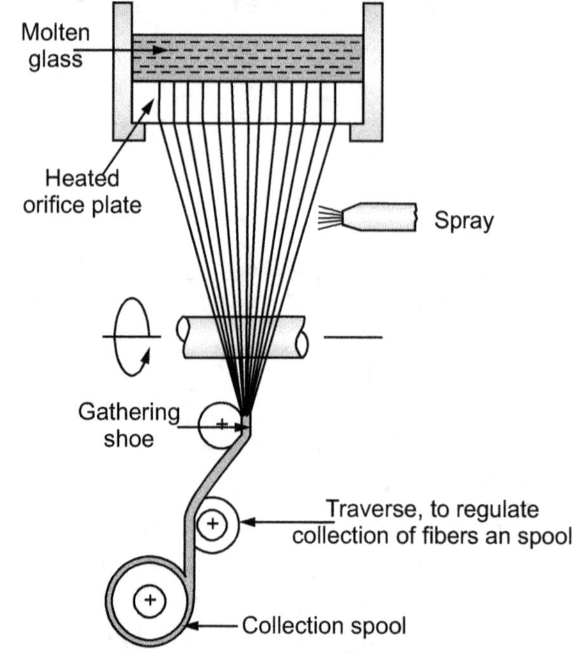

Fig. 5.16 : Drawing of continuous glass fibers

5.10 TECHNIQUES FOR STRENGTHENING AND ANNEALING GLASS

- Glass can be strengthened by a number of processes, and discrete glass products may be subjected to annealing and to other finishing operations to impart desired properties and surface characteristics.

- Thermal Tempering. In this process the surfaces of the hot glass are cooled rapidly by a blast of air.

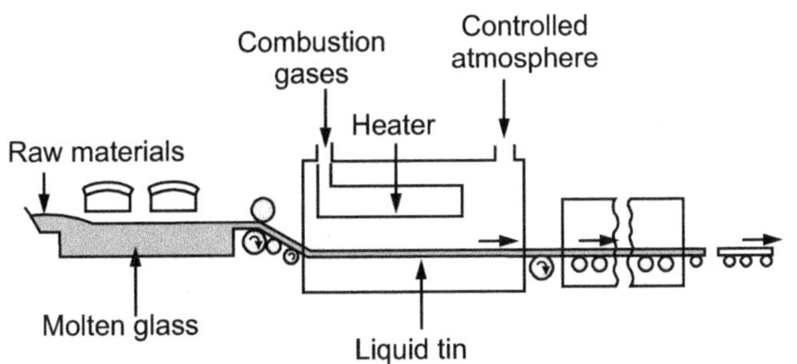

Fig. 5.17 : The stages involved in inducing compressive surface residual stresses for improved strength

- As a result, the surfaces shrink and tensile stresses develop on the surfaces. As the bulk of the glass begins to cool, it contracts.
- The already solidified surfaces of the glass are then forced to contract, and consequently, they develop residual compressive surface stresses, while the interior develops tensile stresses.
- Compressive surface stresses improve the strength of the glass in the same way that they do in metals and other materials.
- The higher the coefficient of thermal expansion of the glass and the lower its thermal conductivity, the higher will be the level of residual stresses developed, and hence, the stronger the glass becomes.
- Thermal tempering takes a relatively short time and can be applied to most glasses. Because of the high amount of energy stored in residual stresses, tempered glass shatters into a large number of pieces when broken. The broken pieces are not as sharp and hazardous as those from ordinary window glass.

5.10.1 Chemical Tempering

- In this process, the glass is heated in a bath of molten KNO_3, K_2SO_4, $NaNO_3$ or depending on the type of glass.
- Ion exchanges then take place, with larger atoms replacing the smaller atoms on the surface of the glass.
- As a result, residual compressive stresses develop on the surface. This condition is similar to that created by forcing a wedge between two bricks in a brick wall.
- The time required for chemical tempering is about one hour longer than that for thermal tempering.
- Chemical tempering may be performed at various temperatures. At low temperatures, part distortion is minimal; therefore, complex shapes can be treated. At elevated temperatures, there may be some distortion of the part, but the product can then be used at higher temperatures without loss of strength.

5.10.2 Laminated Glass

- Laminated glass, a product of another strengthening method called laminate strengthening, it consists of two pieces of flat glass with a thin sheet of tough plastic in between.

- When laminated glass is cracked, its pieces are held together by the plastic sheet a phenomenon commonly observed in a shattered automobile windshield.
- Traditionally, flat glass for glazing windows and doors has been strengthened with wire netting embedded in the glass during its production.
- When a hard object strikes the surface, the glass shatters, but the pieces are held together because of the toughness of the wire, which has both strength and ductility.

5.10.3 Bulletproof Glass

- Laminated glass has considerable ballistic impact resistance and can prevent the full penetration of solid objects because of the presence of a tough polymer film in between the two layers of glass.
- Bulletproof glass (used in some automobiles, armored bank vehicles, and buildings) is a more challenging design, due to the very high speed and energy level of the bullet and the small size and the shape of the bullet tip, representing a small contact area and high localized stresses.
- Depending on the caliber of the weapon, bullet speeds range from about 350 to 950 m/s. Bulletproof glass (also called bullet-resistant glass) ranges in thickness from 7 to 75 mm.
- The thinner plates are designed for resistance to handguns, and thicker plates are for rifles. Although there are several variations, bulletproof glass basically consists of glass laminated with a polymer sheet (usually polycarbonate).
- The capacity of a bulletproof glass to stop a bullet depends on (a) the type and thickness of the glass; (b) the size, shape, weight, and speed of the bullet; and (c) the properties and thickness of the polymer sheet.
- Polycarbonate sheets commonly are used for bulletproof glass. As a material widely used for safety helmets, windshields, and guards for machinery, polycarbonate is a tough and flexible polymer.
- Combined with a thick glass, it can stop a bullet, although the glass develops a circular shattered region. Proper bonding of these layers over the glass surface is also an important consideration, as there usually is more than one round fired during such encounters. Also, in order to maintain the transparency of the glass and minimize distortion, the index of refraction of the glass and the polymer must be nearly identical.
- If a polymer sheet is only on one side of the glass, it is known as a one-way bulletproof glass. In a vehicle, the polymer layer is on the inside surface of the glass.
- An external bullet will not penetrate the window, because the bullet will strike the glass first, shattering

it. The glass absorbs some of the energy of the bullet, thus slowing it down. The remaining energy is dissipated in the polymer sheet, which then stops the bullet. This arrangement allows someone inside the vehicle to fire back.

- A bullet from inside penetrates the polymer sheet and forces the glass to break outwards, allowing the bullet to go through. Thus, a one-way glass stops a bullet fired from outside but allows a bullet to be fired from inside.

5.11 FINISHING OPERATIONS

- As in metal products, residual stresses can develop in glass products if they are not cooled at a sufficiently low rate.
- In order to ensure that the product is free from these stresses, it is annealed by a process similar to the stress-relief annealing of metals.
- The glass is heated to a certain temperature and then cooled gradually. Depending on the size, the thickness, and the type of the glass, annealing times may range from a few minutes to as long as 10 months, as in the case of a 600-mm (24-in.) mirror for a telescope in an observatory.
- In addition to annealing, glass products may be subjected to further operations, such as cutting, drilling, grinding, and polishing. Sharp edges and corners can be smoothed by (a) grinding (as seen in glass tops for desks and shelves) or (b) holding a torch against the edges (fire polishing), which rounds them by localized softening of the glass and surface tension.
- In all finishing operations on glass and other brittle materials, care should be exercised to ensure that there is no surface damage, especially stress raisers such as rough surface finish and scratches.
- Because of notch sensitivity, even a single scratch can cause premature failure of the part, especially if the scratch is in a direction where the tensile stresses are a maximum.

5.12 DESIGN CONSIDERATIONS FOR CERAMICS AND GLASSES

Following Points are important in design of ceramics and glass components :

- Ceramic and glass products require careful selection of composition, processing methods, finishing operations, and methods of assembly with other components.
- The potential consequences of part failure are always a significant factor in designing ceramic and glass products.
- The knowledge of limitations is important such as such as poor tensile strength, sensitivity to internal and external defects, and low impact toughness. On the other hand, these limitations have to be balanced against such desirable characteristics as hardness, scratch resistance, compressive strength at room and elevated temperatures, and a wide range of diverse physical properties.

- The control of processing parameters and of the quality and level of impurities in the raw materials is important. As in all design decisions, there are priorities and limitations, and several factors should be considered simultaneously, including the number of parts needed and the costs of tooling, equipment, and labor.
- Dimensional changes, warping, the possibility of cracking during processing, and service life are significant factors in selecting methods for shaping these materials.
- When a ceramic or glass component is part of a larger assembly, its compatibility with other components is another important consideration.
- Particularly important are the type of loading and thermal the wide range of coefficients of thermal expansion for various metallic and nonmetallic materials. Thus, when a plate glass fits tightly in a metal window frame, temperature variations can cause thermal stresses that may lead to cracking a phenomenon often observed in some tall buildings. A common solution is placing rubber seals between the glass and the frame to avoid stresses due to differences in thermal expansion.
- A general guide is that, in order for a glass item to withstand a load of 1000 hours or longer, the maximum stress that can be applied is about one-third of the maximum stress that it can withstand during the first second of loading.

EXERCISE

1. What are types of ceramics ? Give some application of ceramics.
2. Give the procedure of slip casting process.
3. With a neat sketch explain injection moulding.
4. Enlist different finishing operations. Explain each.
5. Write a note on forming and shaping of glass.
6. Why is "Drying and Firing" technique adopted in ceramic preparation. Explain.
7. Write a shirt note on :
   (i) Slip casting (ii) Tape casting (iii) Pressure casting
8. Draw the tree diagram of processing of ceramics.
9. Explain "Pressing" process of ceramics.
10. Write a note on "Discrete glass products"
11. Explain forming of glass fibers.

6.1 INTRODUCTION

- Plastic is a material consisting of any of a wide range of synthetic or semi-synthetic organics that are malleable and can be molded into solid objects of diverse shapes.
- Plastics are typically organic polymers of high molecular mass, but they often contain other substances.
- They are usually synthetic, most commonly derived from petrochemicals, but many are partially natural

How Plastics are Made?

- Plastics are produced using a process know as polymerization. Polymerization occurs when monomers join together to form long chains of molecules called polymers.
- Polymerization comes from the word 'POLY' which means 'MANY' and 'MER' which means 'PART'. So Polystyrene means 'POLY' many single monomers of 'STYRENE', joined together to form a long chain.

Fig. 6.1: Polymer

Plastics are Made from the Following Materials

- **Plasticizers:** They improve the softening, decrease brittleness and workability of plastics. They are organic substances. They are additives most commonly phthalate ester in PVC.
- **Stabilizers:** They prevent chemical degradation of plastics. By nature they are antioxidants. Hindered amine light stabilizer.
- **Fillers:** They increase the tensile strength of plastics. They are wood flour and glass wool.
- **Reinforcing:** They increase its mechanical strength. Example is glass agents fibre.
- **Pigments:** They are used to impart a particular color to the plastic.

Properties of Plastic

- Light in weight.
- Easy workability.
- High resistant to corrosion.
- Good dimensional stability.
- Absorbent of vibration and sound.
- Good thermal and electrical insulator.
- Good strength and rigidity.
- Impermeable to water.
- Good resistant to chemicals.
- Low fabrication cost.
- Can be made transparent and colored.

6.1.1 Types of Plastic

1. **Thermosetting Plastic: (Heat-Setting Plastic)**

- Thermosetting plastics can only be heated and shaped once. Separate polymers are joined in order to form a huge polymer. The main thermosetting plastics are **epoxy resin, melamine formaldehyde, polyester resin and urea formaldehyde.**

Advantages

- More resistant to high temperatures than thermoplastics.
- Highly flexible design.
- Thick to thin wall capabilities.
- Excellent aesthetic appearance.
- High levels of dimensional stability.
- Cost-effective.

Disadvantages

- Cannot be recycled.
- More difficult to surface finish.
- Cannot be remolded or reshaped.

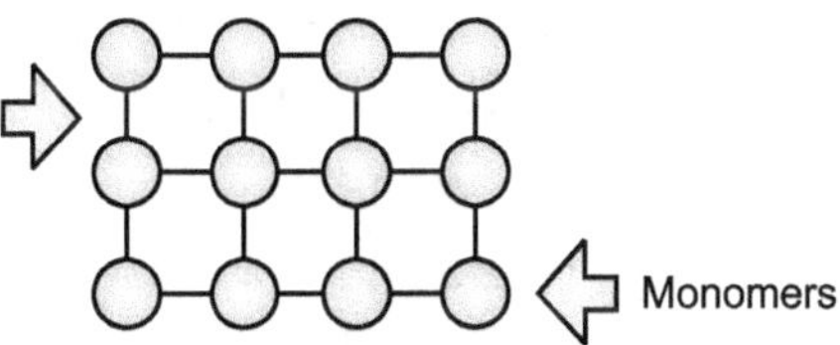

Fig. 6.2: Thermosetting plastic

2. **Thermoplastic: (Recycled Plastic)**

- Thermoplastics pellets soften when heated and become more fluid as additional heat is applied.
- The curing process is completely reversible as no chemical bonding takes place. E.g. Nylon, acrylic, LDPE, HDPE

Advantages

- Highly recycled.
- Aesthetically superior finishes.
- High-impact resistance.
- Remoulding/reshaping capabilities.
- Chemical resistant.
- Hard crystalline or rubbished surface options.
- Eco-friendly manufacturing.

Disadvantages

- Generally more expensive than thermoset.
- Can melt if heated.

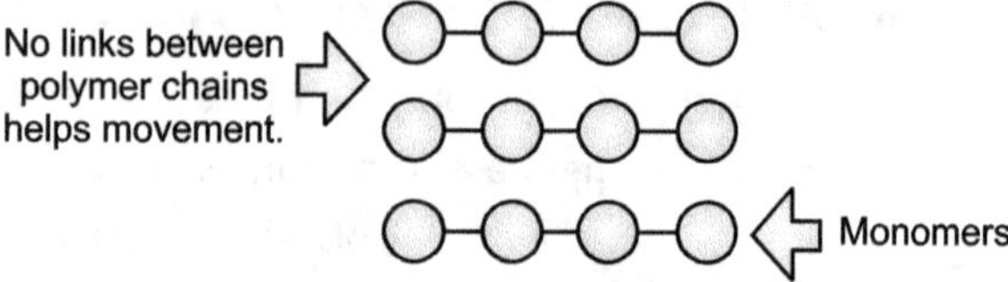

Fig. 6.3 : Thermoplastic

6.2 MOULDING

- Compression moulding.
- Reaction – injection moulding.
- Rotational moulding
- Transfer moulding.
- Injection moulding.
- Blow moulding.
  - Process and equipment.

6.2.1 Compression Moulding   (May 14)

- **Compression Moulding** is a process in which a moulding polymer is squeezed into a preheated mold taking a shape of the mold cavity and performing curing due to heat and pressure applied to the material.

- Compression Moulding cycle time is about 1-6 min, which is longer than Injection Moulding cycle.

- The method is suitable for moulding large flat or moderately curved parts.

  - **Step 1:** Loading a precise amount of moulding compound, called the **_charge_**, into the bottom half of a heated mold;

  - **Step 2:** Bringing the mold half together to compress the charge, forcing it to flow and conform to the shape of the cavity;

  - **Step 3:** Heating the charge by means of the hot mold to polymerize and cure the material into a solidified part; and

- **Step 4:** Opening the mold halves and removing the part from the cavity.

- Materials that are typically manufactured through compression moulding include: Polyester fiberglass resin systems (SMC/BMC), Torlon, Vespel, Poly(p-phenylene sulfide) (PPS), and many grades of PEEK.

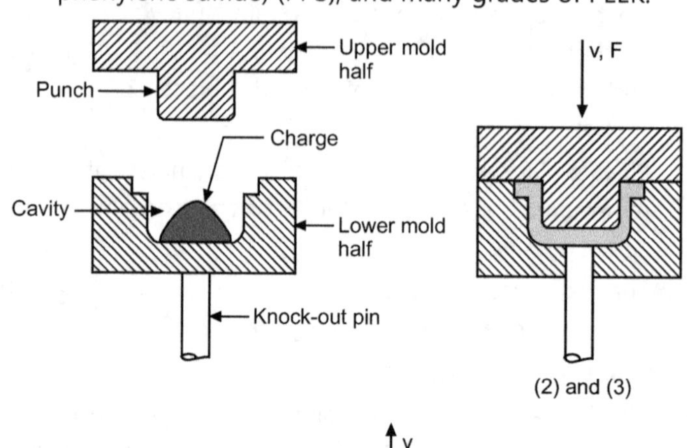

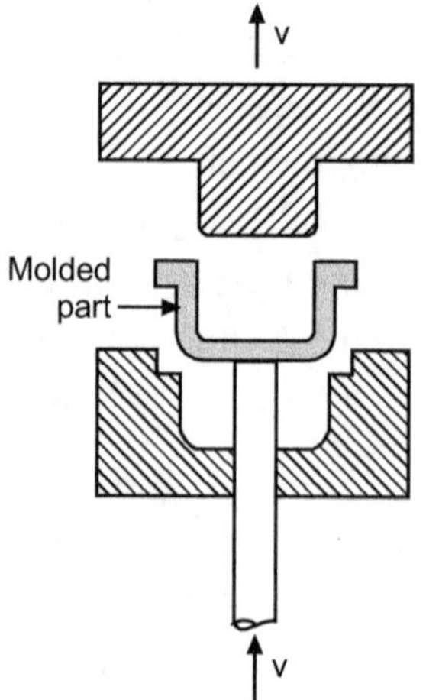

[Compression moulding for thermoplastics: (1) charge is loaded, (2) and (3) charge is compressed and cured, and (4) opening the mould half and removing the part from the cavity.]

Fig. 6.4 : Schematic diagram of Compression moulding

Advantages

- Lowest cost
- More uniform density
- Uniform shrinkage due to uniform flow
- Improved impact strength due to no degradation of fibers during flow
- Dimensional accuracy
- Internal stress and warping are minimized.

Disadvantages

- Curing time large
- Uneven parting lines present
- Scrap cannot be reprocessed.

Applications and Products

- Dinnerware
- Buttons
- Knobs
- Appliance housings
- Radio cases
- Automotive exterior panels especially for commercial vehicles
- Electrical parts.

6.2.2 Transfer Moulding or Gate Moulding

(Dec. 13)

Transfer Moulding

- It is closely related to compression moulding, because it is utilized on the same polymer types (thermosets and elastomers). One can also see similarities to injection moulding, in the way the charge is preheated in a separate chamber and then injected into the mold.

- This type of moulding process, a thermosetting (plastic) charge (preform) is loaded into a chamber immediately ahead of the mold cavity, where it is heated; pressure is then applied to force the softened polymer to flow into the heated mold where curing occurs.

- The method is capable to produce more complicated shapes than compression moulding but not as complicated as injection moulding.

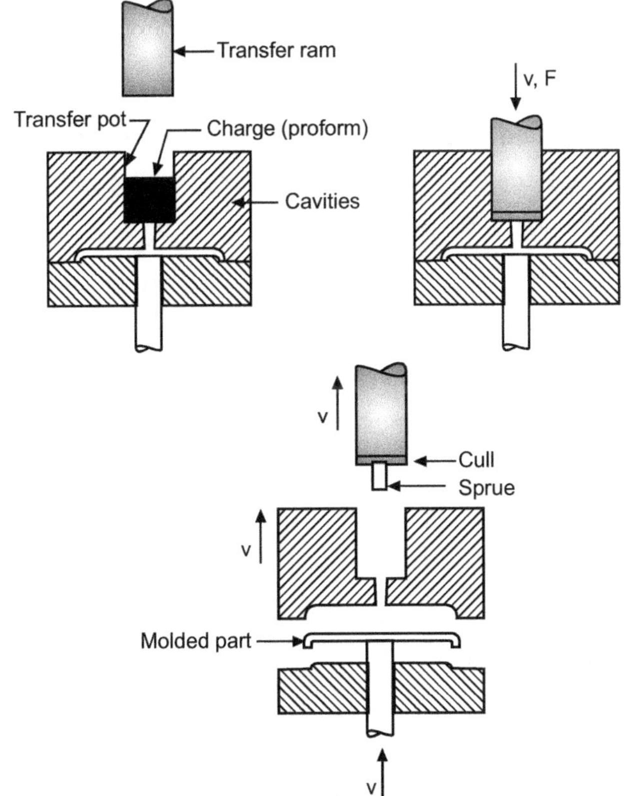

Fig. 6.5 : Schematic diagram of pot transfer moulding

- Transfer moulding is suitable for moulding with ceramic or metallic inserts which are placed in the mold cavity. When the heated polymer fills the mold it forms bonding with the insert surface.

- Transfer moulding is also used for manufacturing radio and television cabinets and car body shells, electronic components with molded terminals, pins, studs, connectors etc.

- Transfer moulding cycle time is shorter than compression moulding cycle but longer than injection moulding cycle.

Advantages of Transfer Moulding

- It can produce parts of better dimensional quality and larger quantities than produced with the compression moulding process.

- Cycle times are generally faster than with compression method.

- Lower unit cost than with compression usually, due to shorter cycle times.

- Less part flashing.

- Since mold is completely closed before the rubber is transferred into the cavity inserted parts are more easily produced.

Limitations of Transfer Moulding

- In both (types) cases of transfer moulding, scrap is produced each cycle in the form of the leftover material in the base of the well and lateral channels, called the '**cull**'.

- In addition, the sprue in pot transfer is scrap material. Because the polymers are thermosetting, the scrap cannot be recovered.

Applications of Transfer Moulding

- Transfer moulding is capable of moulding part shapes that are more intricate than compression moulding but not as intricate as injection moulding.

- In the semiconductor industry, package encapsulation is usually done with transfer moulding due to the high accuracy of transfer moulding tooling and low cycle time of the process.

- Some common products are utensil handles, electric appliance parts, electronic component, and connectors.

- Transfer moulding is widely used to enclose or encapsulate items such as coils, integrated circuits, plugs, connectors, and other components.

6.2.3 Injection Moulding   (May, Dec. 14)

- Injection moulding is a manufacturing process for producing parts by injecting material into a mold.

- Powder or granules from a hopper into a steel barrel with a rotating screw.

- The barrel is surrounded by heaters The screw is forced back as plastic collects at the end of the barrel.

- A sufficient charge of melted plastic has accumulated a hydraulic ram forces the screw forward injecting the thermoplastic through a sprue into the mould cavity.

- Pressure is kept on the mould until the plastic has cooled sufficiently for the mould to be opened and the component ejected.

- Normally thermoplastics are used in this process although a few thermosetting plastics can also be injection moulded.

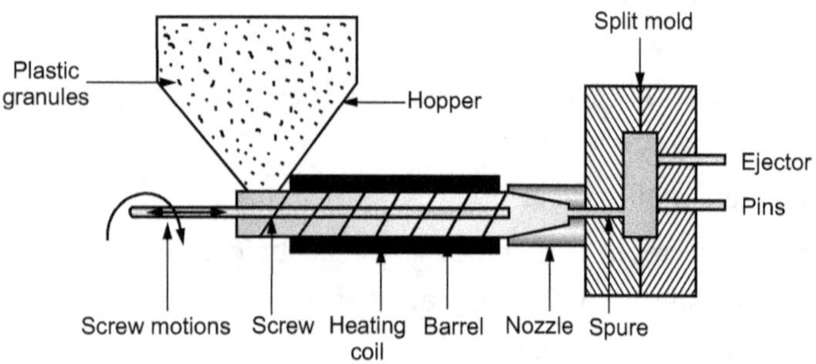

Fig. 6.6 : Injection moulding

Advantages of Injection Moulding

- More precise control of material mixing and flow resulting in better part quality and consistency.

- Fewer cavities are used in this process compared to transfer moulding resulting in less dimensional variation among parts produced from a given mold.

- Lower unit cost (for larger order quantities) due to faster curing and cycle times and higher production rates.

- Less reject material (vs. the transfer process where some material is lost in the transfer pot).

- Is more cost effective when moulding into metal inserts.

- Molds may be less expensive since they are not making a separate ram and pot.

Limitations of Injection Moulding

- Injection moulding is a complex technology with possible production problems.

- Limitations can be caused either by defects in the molds, or more oftened by the moulding process itself.

- When filling a new or unfamiliar mold for the first time, where shot size for that mold is unknown, a technician/tool setter may perform a trial run before a full production run.

- The initial cost is high; however the per-piece cost is low, so with greater quantities the unit price decreases.

Applications of Injection Moulding

- Injection moulding is used for manufacturing DVDs, pipe fittings, battery casings, automobile bumpers and dash boards, television cabinets, electrical switches, telephone handsets, automotive power brake, electrical parts, etc.

6.2.4 Blow Moulding   (Dec. 13)

The principle of blow moulding

- A simple explanation of the principle of blow moulding is similar to inflating a balloon.

Fig. 6.7 : Principle of blow moulding

Blow Moulding is a process in which a heated hollow thermoplastic tube (**parison**) is inflated into a closed mold conforming the shape of the mold cavity.

- Parison - in blow moulding, the hollow tube of plastic melt extruded from the die head, and expanded within the mold cavity by air pressure to produce the molded part

- The blow moulding process begins with melting down the plastic and forming it into a heated hollow thermoplastic tube (parison or perform).

- The parison is a tube-like piece of plastic with a hole in one end through which compressed air can pass.

- The air pressure then pushes the plastic out to match the mold. Once the plastic has cooled and hardened the mold opens up and the part is ejected.

6.2.5 Extrusion Blow Moulding   (Dec. 13)

Extrusion Blow Moulding involves manufacture of parison by conventional extrusion method using a die similar to that used for extrusion pipes.

* **Step 1:** The parison is extruded vertically in downward direction between two mold halves.

* **Step 2:** When the parison reaches the required length the two mold halves close resulting in punching the top of parison end and sealing the blow pin in the bottom of the parison end.

* **Step 3:** Parison is inflated by air blown through the blow pin, taking a shape conformed that of the mold cavity. The parison is then cut on the top.

* **Step 4:** The mold cools down, its halves open, and the final part is removed.

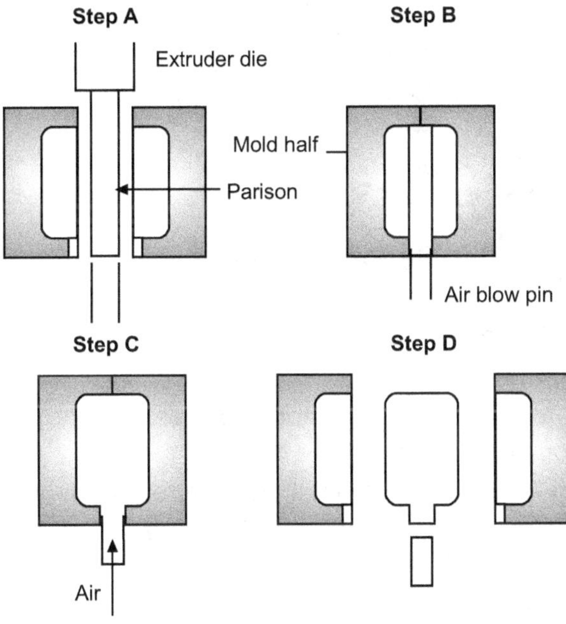

Fig. 6.8 : Extrusion blow moulding

* Extrusion blow moulding is commonly used for mass production of plastic bottles. The production cycle consists of the following steps

* The most widely used materials for blow moulding are: Low Density Polyethylene (LDPE), High Density Polyethylene (HDPE), Polypropylene (PP), Polyvinyl Chloride (PVC), Polyethylene Terephtalate (PET).

* Disposable containers of various sizes and shapes, drums, recyclable bottles, automotive fuel tanks, storage tanks, globe light fixtures, toys, tubs, small boats are produced by Blow moulding method.

* There are three principal techniques of Blow Moulding, differing in the method by which parisons are prepared

are 1. Extrusion blow moulding, 2. Injection blow moulding, 3. Stretch blow moulding.

Advantages of Blow Moulding

* Advantages of blow moulding include: low tool and die cost; fast production rates; ability to mold complex part; produces recyclable parts.

Limitations of Blow Moulding

* Limitations of blow moulding include: limited to hollow parts, wall thickness is hard to control.

6.2.6 Reaction-Injection Moulding (RIM)

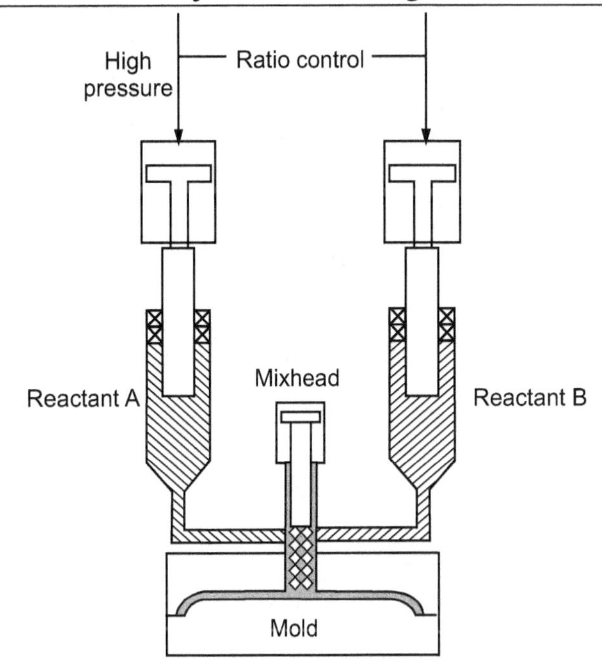

Fig. 6.9 : Reaction injection moulding

RIM Process and Equipment

* Day tanks

  ➢ Store unfilled components and filled (reinforced) components

  ➢ Slowly and continuously agitate by mixing impellers to prevent settling of fibers

  ➢ May have heating jackets for temp control

  ➢ Pressurized at 15 - 60 psi

* Metering cylinders fill with components from day tanks

  ➢ Components could be moved directly from the pressurized tanks with air pressure, however usually transferred with the assistance of feeder pumps feeder pumps can provide continuous recirculation (maintain filler in suspension and uniform temp)

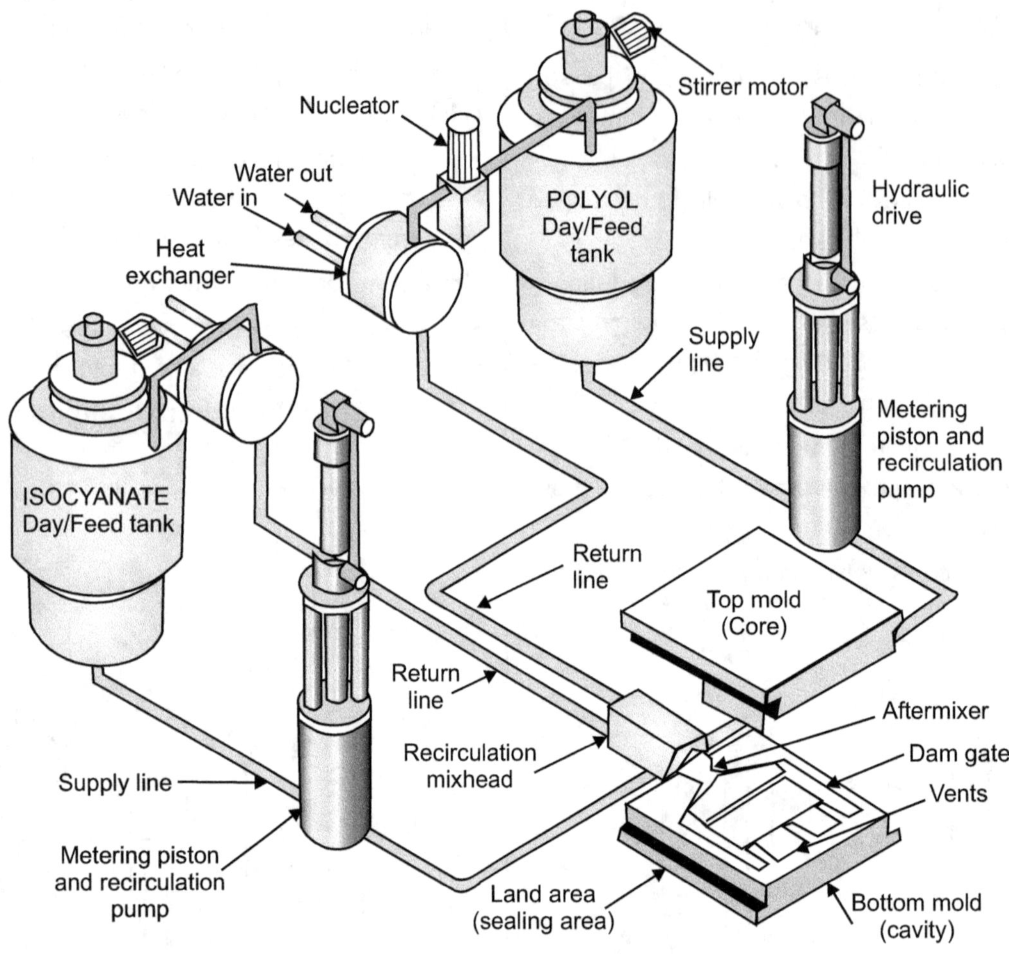

Fig. 6.10 : Schematic of RIM Process

Mixing of Components

- After metering cylinders filled, they are driven forward by hydraulic pressure to deliver components to mixhead at a known rate (delivery of the two components must be closely synchronized to ensure a uniform reaction and consistent properties in cured part)

- During first part of plunger travel, valve to mixing chamber is closed - components are recirculated through head into return lines and back to day tanks

- After preliminary recirculation - mixhead valve is opened, components enter the chamber to be mixed

- Mix chamber is usually small cylinder - components enter from opposite sides of chamber

- Mixhead is designed to develop turbulence in the mix chamber to intimately mix the two components

- Turbulence created by stream impingement at high pressure (1,500 - 3,000 psi)

- Streams should have equal momentum at the time they meet

Filling the Mold

- After metering the shot, mixhead valve is closed, components recirculated back into day tanks

- Metered shot is cleared from mixing chamber by a close fitting ram and flows directly into mold (no solvent flush required)

- Mold normally filled from bottom so air can easily push out ahead of flow

- Shot fills mold to about 90% and expansion during chemical reaction of the polyol and isocyanate completes the fill

Curing and Demoulding

- Components react and gel within 2 - 10 secs from the start of injection

- Mold remains closed for a period of time to allow sufficient cure so part can be removed and handled without damage (30 - 90 secs)

- Knockout pins, automatic slides, or pneumatic devices in the mold used to assist with demoulding

- For small flexible parts may use rubber spatula to pry part loose or insert air nozzle and blow off

- For most resin formulations, parts are post-cured in an oven

Advantages/Disadvantages of RIM Process :

- RIM resin builds viscosity rapidly (higher average viscosity during mold filling)
  - Applications must be simple geometries
  - SRIM preform must be less complex and lower in reinforcement content
  - Parts do not normally flash out of mold parting line sufficiently to require sealing beyond metal land area or a pinch off around perimeter of part (low viscosity of RTM resin requires gasket or o-ring)
- Highly reactive nature of RIM resin systems leads to cycle times currently faster than achieved with RTM process
- Mix ratios of RIM resin systems nearly 1:1 in volume
  - Ideally suited to impingement mixing process

6.2.7 Rotational Moulding

- Rotational moulding is a process of making hollow articles.
- The part is formed inside a closed female mould.
- In this process the mould rotates biaxially during heating and cooling cycle.
- Rotational moulded pieces are stress free because the pieces are produced without any external pressure.
- The ability to manufacture large containers of capacity 30,000 gallons as well as small items like golf ball is responsible for the growth of this process.
- The Process requires relatively in expensive equipment and exerts on only small pressure on the material being formed.

Principle

- The principle of the process is that finely divided plastic material becomes molten when comes in contact with hot metal surface of the mould and takes up the shape of that surface.
- As only female mould is used, the only pressure exerted are those induced by gravity and centrifugal force.
- The polymer is then cooled while still in contact with the metal mould to get the solid copy of the surface.
- Rotational moulding permits to make a wide variety of fully and partially closed items.

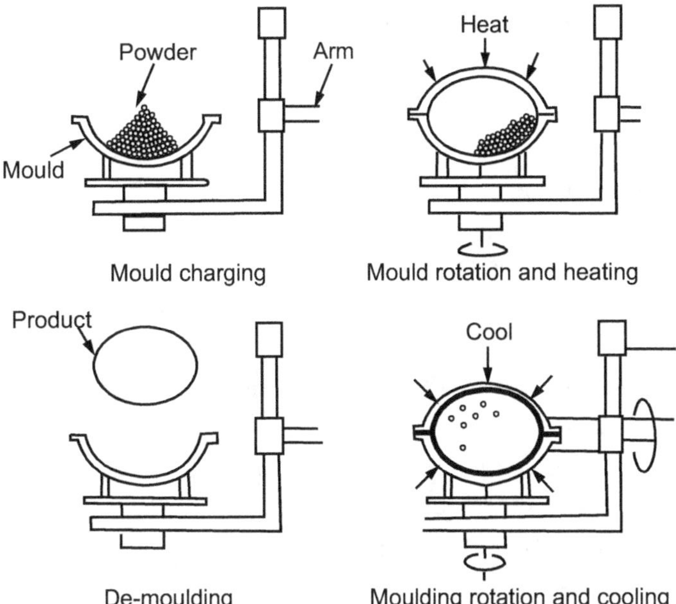

Fig. 6.11 : Principle of rotational moulding

Fig. 6.12 : Rotational moulding components

Advantages

- The major advantage of rotational moulding as compared to other plastic moulding processes is that it can make very large parts.
- It requires comparatively low cost input.
- The products are stress free with strong out side corners. There are no weldlines, sprue or gate marks.
- Here impact toughness is improved and failure due to brittleness is reduced
- The external dimensional details can be easily moulded with better surface glossiness .
- The colour changes in the product can be made easily. Similarly mould changes can also be done rapidly.
- Multilayer moulding is also possible for providing chemical resistance and strength to the part.
- Good control over wall thickness variation is also achievable as compared to blow moulding or thermoforming.

- Moulding can be done with metal inserts and minor undercuts.
- No scrap or very little scrap is produced.
- Low tooling cost.

Disadvantages

- The moulding cycles are longer compared to blow moulding and thermoforming.
- In case of big parts loading and unloading is very labour intensive.
- The process is not suitable for parts with wall thickness less than 0.03".
- The conversion of plastic granules to powder form increases the equipment and process cost.

Rotational Moulding Process

- Loading
- Heating & Moulding
- Cooling
- Unloading

Loading

- This step includes weighing of the charge for a particular product then transferring it to the open cold mould.
- The mould surface usually coated with a mould releasing agent.
- The raw material can be in the form of powder or liquid state.
- The wall thickness can be controlled by varying the amount of raw material charged.
- After the material is charged the mould is closed and clamped to the arm of the machine.
- Then the mould is moved to an oven for heating

Heating and Moulding

- The mould fixed to the arm now moved to a closed chamber where it undergoes intense heating.
- During heating the mould rotates in two planes perpendicular to each other.
- The rotational speed varies in the range of 0-40 rpm on minor & 0-12 rpm on the major axis.
- 4:1 ratio is the most commonly used for symmetric article.
- For moulding unsymmetrical products a wide variability of ratios is necessary.
- The revolving motion distributes the plastic material uniformly over the inside surface of the mould.

- The plastic material fuses into layers to form a hollow article.
- In case of hot air oven the temperature should be between 200°c to 500°c.

Cooling

- For cooling the mould is transferred to the cooling station while still rotating.
- The cooling should be made as quickly as possible to avoid the plastic part to shrink away form the mould.
- Otherwise the part will get distorted.
- Cooling can be done by air or water. To provide faster cooling cold water is sprayed over the mould.

Unloading

- After cooling the mould is transferred to the unloading station.
- In this step the mould is opened and the cooled part is taken out.
- It can be done manually or with mechanical assistance.
- The ejection can also be done by forced air.
- The mould is cleaned and the charge is loaded for the next cycle.

6.3 EXTRUSION OF PLASTICS   (May 14)

- Raw materials in the form if thermoplastic pallets, granules, or powder, placed into a hopper and fed into extruder barrel.
- The barrel is equipped with a screw that blends the pallets and conveys them down the barrel
- Heaters around the extruder's barrels heats the pellets and liquefies them
- Pellets: Extruded product is a small-diameter rod which is chopped into small pellets

Screw has 3-Sections

- Feed section
- Melt or transition section
- Pumping section.
- Most screws have these three zones
  - ➤ **Feed Zone:** Also called solids conveying. This zone feeds the resin into the extruder, and the channel depth is usually the same throughout the zone.
  - ➤ **Melting Zone:** Also called the transition or compression zone. Most of the resin is melted in this section, and the channel depth gets progressively smaller.

> **Metering Zone:** Also called melt conveying. This zone, in which channel depth is again the same throughout the zone, melts the last particles and mixes to a uniform temperature and composition.

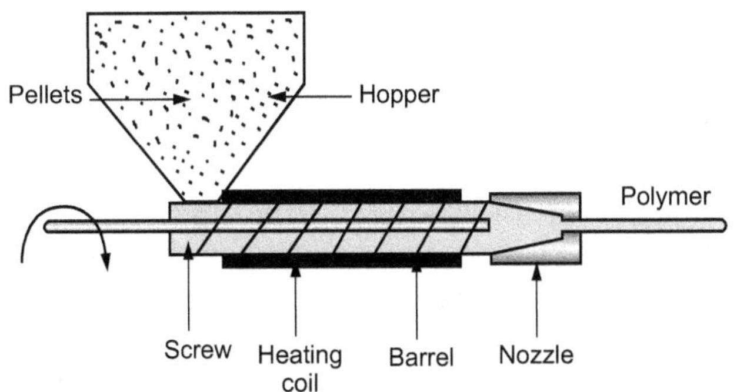

Fig. 6.13 : Extrusion of plastics

- Complex shapes with constant cross-section.

- Solid rods, channels, tubing, pipe, window frames, architectural components can be extruded due to continuous supply and flow.

- Plastic coated electrical wire, cable, and strips are also extruded.

Advantages of Plastics Extrusion Process

- The production capabilities of plastic extrusion technology allow you to manufacture high volumes.

- Plastic extrusion is extremely versatile.

- It allows you to create products of complex shapes with different sizes, thicknesses, hardnesses, textures and colours.

- The co-extrusion process enables manufacturers to incorporate different plastic materials and compounds into one product.

- Multiple layers are created which are then 'fused' into one.

Limitations of Plastics Extrusion Process

- Uneven flow at this stage would produce a product with unwanted stresses at certain points in the profile. These stresses can cause warping upon cooling.

- Limited complexity of parts.

- Uniform cross-sectional shape only.

Applications of Plastics Extrusion Process

- Thin film (flat or tubular) is the most common product.

- Other extruded products include pipe and tubing, coated paper or foil, monofilaments and textile fibers, flat sheet, wire and cable covering etc.

6.3.1 Film Extrusion

Film: thickness < 0.5 (<0.25 mm) mm for packaging purpose

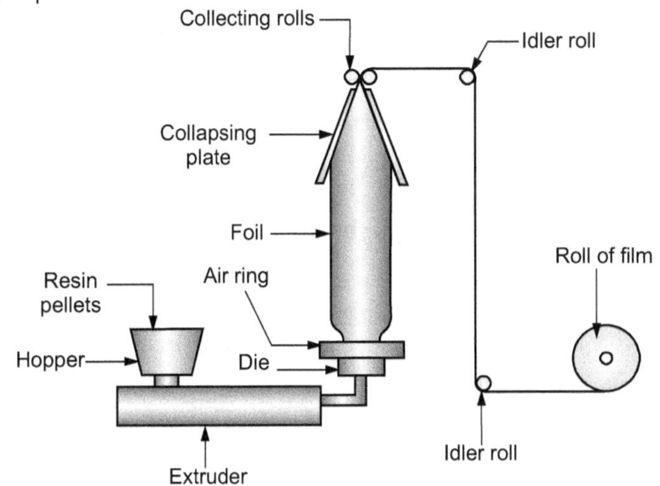

Fig. 6.14 : Film extrusion

- Combine principle of extrusion and blowing to produce film.

- Packaging application like product wrapping material grocery bags, garbage bags, etc.

- It is complex process which combines the principle of extrusion and blowing to produce a thin film.

- Air pressure in the bubble must be kept constant to maintain uniform thickness.

- Nip roll and idler roll are used to restrain the blow tube.

6.3.2 Pipe Extrusion

- Extruded tubing process, such as drinking straws and medical tubing, is manufactured the same as a regular extrusion process up until the die.

- Hollow sections are usually extruded by placing a pin or mandrel inside of the die, and in most cases positive pressure is applied to the internal cavities through the pin.

- Tubing with multiple lumens (holes) must be made for special applications. For these applications, the tooling is made by placing more than one pin in the center of the die, to produce the number of lumens necessary.

- In most cases, these pins are supplied with air pressure from different sources. In this way, the individual lumen sizes can be adjusted by adjusting the pressure to the individual pins.

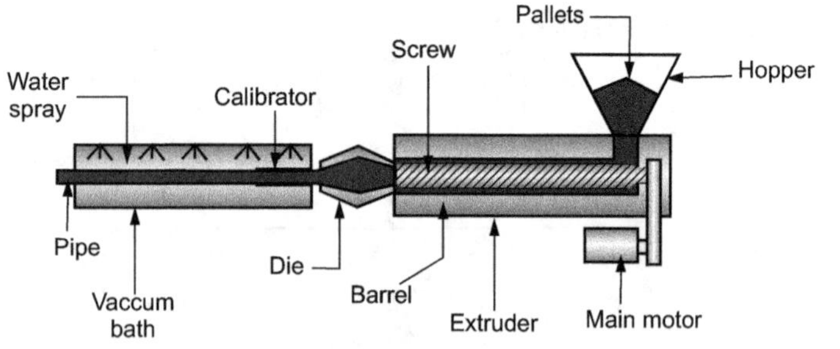

Fig. 6.15 : Pipe extrusion

6.3.3 Wire/Cable Extrusion

- In a wire coating process, bare wire (or bundles of jacketed wires, filaments, etc.) is pulled through the center of a die similar to a tubing die.

- There are two different types of extrusion tooling used for coating over a wire. They are referred to as either "pressure" or "jacketing" tooling.

- If intimate contact or adhesion is required, 'pressure tooling' is used. If it is not desired, 'jacketing tooling' is chosen.

- When the bare wire is fed through the pin, it does not come in direct contact with the molten polymer until it leaves the die.

- For pressure tooling, the end of the pin is retracted inside the crosshead, where it comes in contact with the polymer at a much higher pressure.

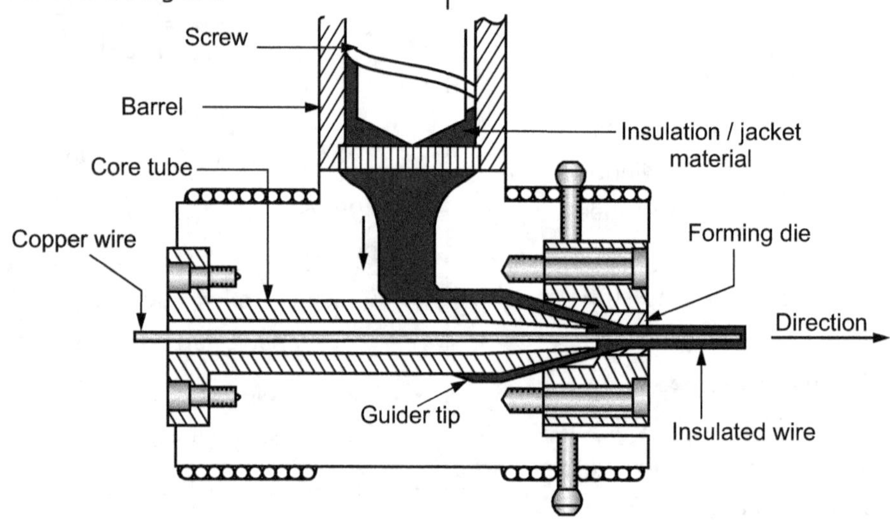

Fig. 6.16 : Wire extrusion

6.3.4 Sheet Extrusion-Calendaring

- Calendaring is a specialist process for high-volume, high quality plastic film and sheet, mainly used for PVC as well as for certain other modified thermoplastics.

- The melted polymer is subject to heat and pressure in an extruder and formed into sheet or film by calendaring rolls.

- The temperature and speed of the rolls influences the properties of the film.

- Calendaring allows special surface treatments of the film or sheet such as embossing or enhancing the physical properties or in-line lamination.

6.4 THERMOFORMING

- **Thermoforming** is a process in which a flat thermoplastic sheet is heated and deformed into the desired shape.

  The process is widely used in packaging of consumer products and to fabricate large products such as bathtubs, contoured skylights, and internal door liners for refrigerators.

- **Thermoforming** is a process of shaping flat thermoplastic sheet which includes two stages: softening the sheet by heating, followed by forming it in the mold cavity.

  There are three thermoforming methods, differing in the technique used for the forming stage

  ➢ Vacuum Thermoforming

  ➢ Pressure Thermoforming

  ➢ Mechanical Thermoforming.

6.4.1 Vacuum Thermoforming   (May 14)

- In case of thermoforming process, a negative pressure is used to draw a preheated sheet into a mold cavity. The process is explained using schematic diagram shown in Fig. 6.17. The holes for drawing the vacuum in the mold are on the order of 0.8 mm in diameter, so their effect on the plastic surface is minor.

- Referring Fig. 6.17 of vacuum thermoforming;

  Step 1: A flat plastic sheet is softened by heating

  Step 2: The softened sheet is placed over a concave mold cavity;

Step 3: A vacuum draws the sheet into the cavity; and

Step 4: The plastic hardens on contact with the cold mold surface, and the part is removed and subsequently trimmed from the web.

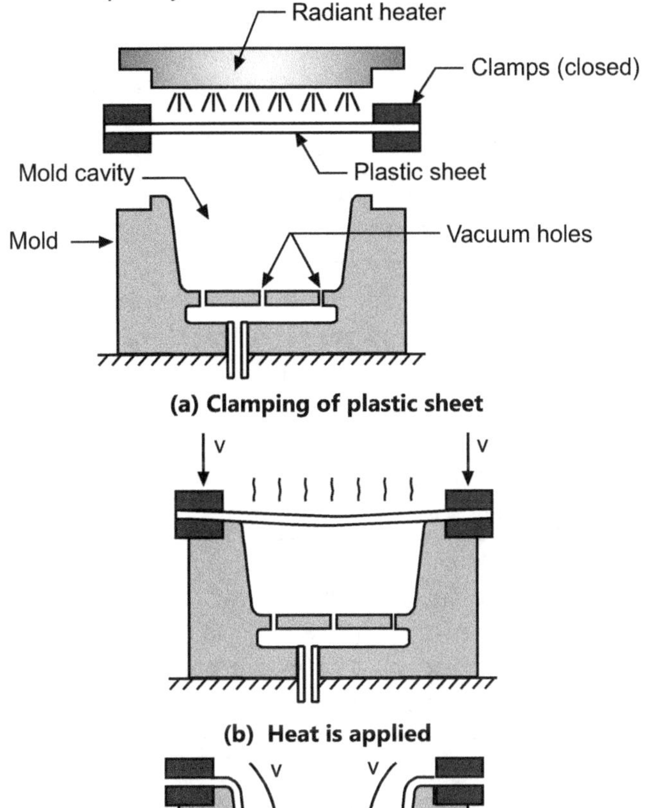

(a) Clamping of plastic sheet

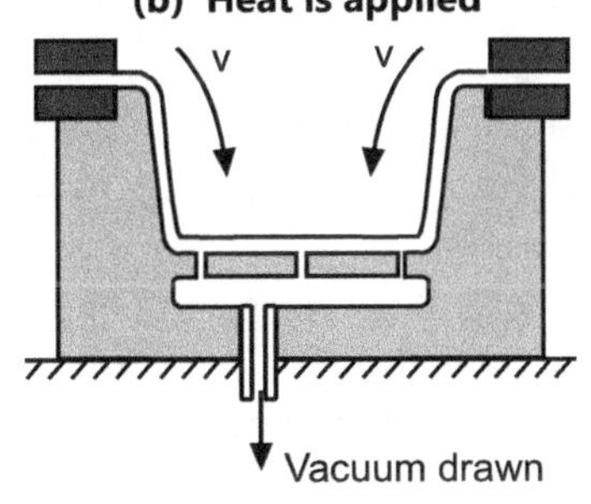

(b) Heat is applied

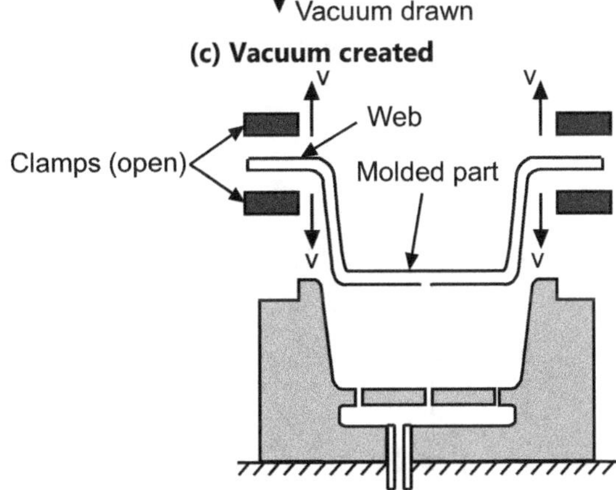

(c) Vacuum created

(d) Unclamping of plastic sheet

Fig. 6.17 : Vacuum-thermoforming

6.4.2 Pressure Thermoforming

- The process involves shaping a preheated thermoplastic sheet by means of air pressure.
- The air pressure forces the soft sheet to deform in conformity with the cavity shape.

- When the plastic sheet comes into the contact with the mold surface it cools down and hardened.

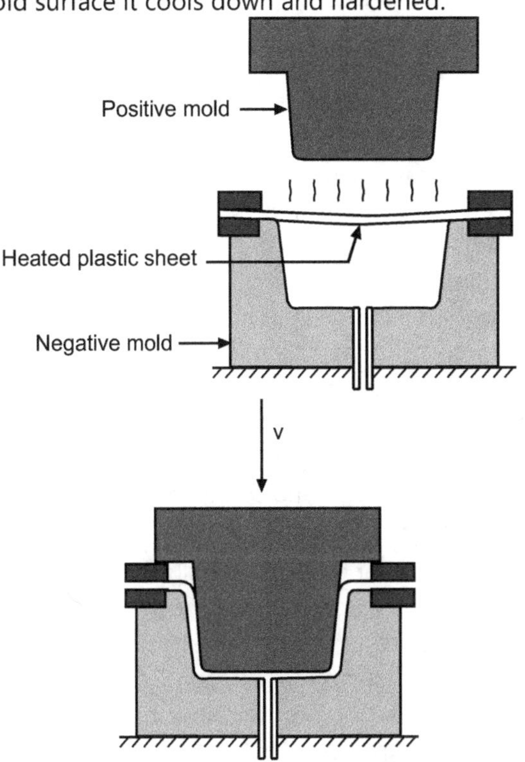

Fig. 6.18 : Pressure-Thermoforming

- In the case of the positive mold, the heated sheet is draped over the convex form and negative or positive pressure is used to force the plastic against the mold surface. The positive mold is shown in below for the case of vacuum forming.

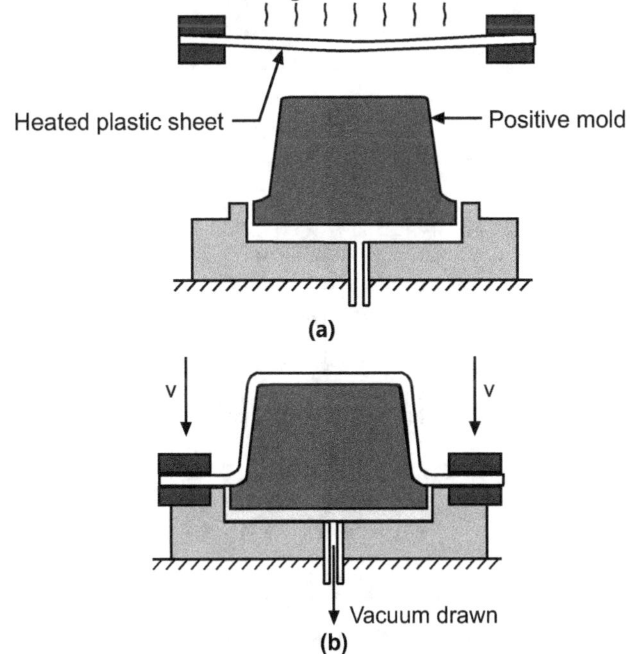

Fig. 6.19 : Pressure-Thermoforming with positive mold

Advantages of Thermoforming Process

- Its advantages are better dimensional control,
- Better opportunity for surface detailing on both sides of the part.

Limitations of Thermoforming Process
- Main limitation is that two mold halves are required,
- The molds for the other two methods are therefore less costly.
- Only thermoplastics can be thermoformed, since extruded sheets of thermosetting or elastomeric polymers have already been cross-linked and cannot be softened by reheating.

Applications of Thermoforming Process
- Thin film packaging items that are mass produced by thermoforming include blister packs and skin packs.
- Thermoforming applications include large parts that can be produced from thicker sheet stock. Examples include covers for business machines, boat hulls, shower stalls, diffusers for lights, advertising displays and signs, bathtubs, and certain toys.

6.4.3 Mechanical Thermoforming

- It is similar to compression moulding.
- It uses matching positive and negative molds that are brought together against the heated plastic sheet, forcing it to assume their shape.
- In the pure mechanical forming method, air pressure (positive or negative) is not used at all.
- The forming sequence (steps 1 and 2) are : (1) sheet is placed over a mold cavity; and (3) positive mechanically forces the sheet into the cavity, which is illustrated in Fig. 6.20.

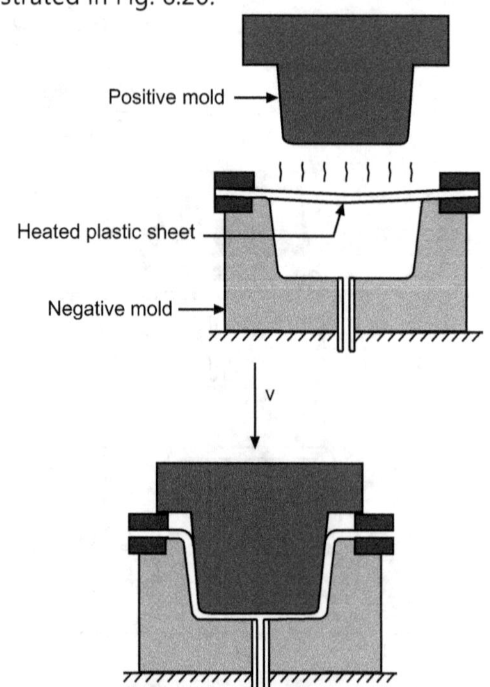

Fig. 6.20 : Mechanical-Thermoforming

Advantages of Mechanical Thermoforming Process
- Its advantages are better dimensional control,
- Better opportunity for surface detailing on both sides of the part.

Limitations of Mechanical Thermoforming Process
- The molds for the other two methods are therefore less costly.
- Main limitation is that two mold halves are required,

Applications of Mechanical Thermoforming Process
- Thermoforming is a secondary shaping process.
- Mass production thermoforming operations are performed in the packaging industry.
- The operations are often designed to produce multiple parts with each stroke of the press using molds with multiple punches and cavities.
- Thin film packaging items that are mass produced by thermoforming include blister packs and skin packs.
- Thermoforming applications include large parts that can be produced from thicker sheet stock.
- Examples include covers for business machines, boat hulls, shower stalls, diffusers for lights, advertising displays and signs, bathtubs, and certain toys.

6.5 CASTING

- Simple, inexpensive but slow.
- Typical parts: gears, bearings, wheels, thick sheets.
- Convention casting of TP: a mixture of monomer, catalyst, and various additives is heated and poured into the mold. Part forms after polymerization takes place at ambient pressure.
- Centrifugal casting
- **Potting and Encapsulation:** Casting plastic material around electric component [transformer]. Potting is done in a housing or case, which is an integral part of product. In encapsulation, component is coated with a layer of the solidified plastic.

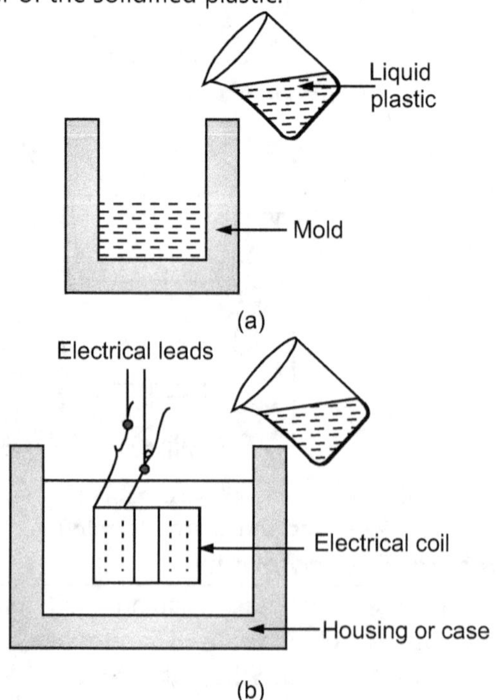

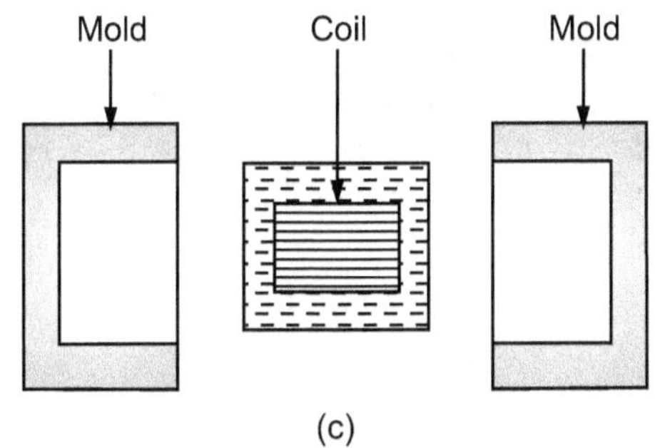

Fig. 6.21 : Schematic illustration of (a) Casting, (b) potting, and (c) encapsulation processes for plastics and electrical assemblies, where the surrounding plastic serves as a dielectric

6.6 FOAM MOULDING

- Products are styrofoam cups, food containers, insulating blocks and shaped packaging materials.
- In foam moulding, raw material is expandable polystyrene beads where products have a cellular structure.
- Structure may have open and interconnected porosity or have closed cells.
- Amount of expansion can be controlled by varying the temperature and time.
- A common method of foam moulding is to use pre-expanded polystyrene beads.

Structural Foam Moulding (SFM)

- SFM process is used to make plastic products with a solid outer skin and a cellular inner structure.
- Typical Products: furniture components, TV cabinets, business machine housings.

Injection Foam Moulding:

- thermoplastics are mixed with a blowing agent (inert gas such as N_2), which expands the material.
- The core of part is cellular, and the skin is rigid.
- Thickness of skin: up to 2 mm.
- Part densities as low as 40% of the density of solid plastic.
- Polyurethane foam processing: it starts with mixing of two or more chemical components, the reaction forms cellular structure which solidifies in the mold.

6.6.1 Cold Forming

Cold Metal Working Processes can be Applied to Many Thermoplastics

- Materials Rolling, deep drawing, extrusion, closed die forging...

Typical Materials

- Polypropylene, polycarbonate, ABS, PVC

Necessary Attributes for Cold Forming

- Ductile at room temperature
- Deformations must be non recoverable (minimize spring back and creep) Cold forming

Cold Metal Working Processes can be Applied to Many Thermoplastics

- Materials Rolling, deep drawing, extrusion, closed die forging...

Typical Materials

- Polypropylene, polycarbonate, ABS, PVC

Necessary Attributes for Cold Forming

- Ductile at room temperature
- Deformations must be non recoverable (minimize springback and creep)

6.6.2 Solid-Phase Forming

- At temperatures 10-20 °C lower than the melting temperature
- Springback is lower than for cold forming

6.7 PROCESSING ELASTOMERS

Calendaring

- Calendaring is the process of squeezing a plastic melt between two or more counter rotating cylinders or rolls to form a continuous film and sheet
- Both rigid and plasticized compounds of PVC resin can be used in calendaring process.
- The sheet is formed in the range of 0.1 to 1.0 mm and above thickness. The sheet can be transparent, colored, embossed, printed or laminated form.
- The PVC resin is generally processed in calendaring plant, and it needs blending, mixing and compounding by addition of suitable chemicals called as additives and fillers to make compound before processing in the calendaring plant.
- Other then PVC resin ABS, PE, PP and Styrene are also used for calendaring process.
- PVC resin for rigid compounds made either by the suspension or bulk polymerization process is best in the 1.7 to 2.0 relative viscosity range.
- Homopolymer grades are used alone and in combination with 2 to 10 percent acetate copolymer.
- Heat stabilizers are needed, when heated PVC naturally tends to degrade, first by yellowing then by turning dark brown and losing its physical properties.
- Many types of stabilizers are used, Metallic salts, mixed metal salts, organotins and tin mercaptides are the major categories.
- Impact modifiers are used to absorb shock to improve impact resistance of PVC compound. The selection of ABS, MBS, Chlorinated polyethylene(CPE) or acrylic polymer chosen on the basis of their impact efficiency, clarity, weather ability and stress whitening, as well as their processing characteristics

- Process aids assist stabilization and increase the melt strength of the sheet during calendaring and post processes such as thermoforming.
- The pigments or dyes which can withstand temperatures of processing at about 180°C should be considered. Pigments which are suitable for coloring plastic materials are metallic salts, lead, Cadmium based and also organic colors are used.

Calendaring Process

- The rolls are heated either by steam or hot oil and roll temperature ranges from 150° to 180°C.
- Fluxed material delivered to the first calendar nip is regulated to form a rolling bank. The sheet passing the first nip forms another bank between the second and third rolls, at the final nip, the desired thickness is obtained from the smallest bank possible to minimize stress in the sheet.

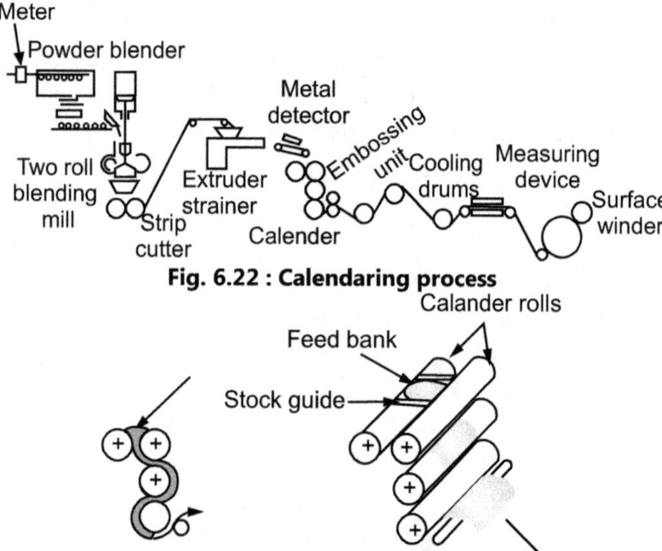

Fig. 6.22 : Calendaring process

Fig. 6.23 : Calendaring process

- After fluxing and gelation, either a two roll mill or an extruder is used to form strands or any acceptable forms.
- Melt is maintained and delivered to the calendar at around 140-160°C. The calendar roll squeezes a plastic melt into a flat sheet.
- The sheet is cooled by passing it over cooling rolls then through a thickness gauge (e.g. betray device) prior to being wound-up.
- If the sheet is wound up too hot, it has a tendency to shrink during the cooling process.
- The sheet is then trimmed to required width and wound on a tube or cut to length and stacked as sheets.
- The trimmed edges of the sheet can be recycled.

Advantages of Calendaring Process

- Delivery/ output of a calendar is high
- Versatile in product design and pattern and permits lamination and embossing without additional equipment.
- Increase in maximum width of film / sheet.
- Better control of film / sheet thickness.
- The properties are more uniform across the width of the product.
- Better optical properties due to cooling chilled rolls.
- PVC being heat sensitive material, it is safer by calendaring process than extrusion process.

Applications

- The applications include sheets in hospitals, Printer blankets, rain coats, table cloths, draperies, water tank liners, wall covers, thermoform sheets, packaging applications, medical applications, stationery items, files, folders, luggage bag applications, counter tops, flooring, upholstery, automobiles applications, and Toys.
- They may be heat-sealed, heat shrunk for blister packaging or used plain as wrapping materials, mulches, swimming pool liners, reservoir films, vapor barriers.

EXERCISE

1. Compare thermoplastic and thermosetting plastics.
2. Explain the construction and working of compression moulding process and equipment.
3. Write short note on
   (i) Transfer moulding,
   (ii) Blow moulding,
   (iii) Injection moulding.
4. Explain the construction and working of Injection moulding process and equipment.
5. Explain working principle of screw type injection moulding process.
6. Explain the construction and working of extrusion of plastic.
7. Discuss the type of extruder.
8. Explain with neat diagram how extruders are specified?
9. Discuss how pipe, cable and sheet are produced?
10. List thermoforming process and explain any one in detail.
11. Compare pressure forming and vacuum forming.
12. Explain calendaring process with neat sketch.
13. What is pressure thermoforming.

Model Question Papers for
End-Semester Examination

PAPER I

Time : 3 Hours   **Max. Marks : 60**

Instructions to the candidates :

    (1) Each Question carries 12 Marks.

    (2) Attempt any five questions to the following.

    (3) Illustrate your answers with neat sketches, diagram etc., wherever necessary.

    (4) If some part or parameter is noticed to be missing, you may appropriately assume it and should mention it clearly.

1. **(a)** For rough grinding operations, determine the machining time required when cutting speed is 25m/min, Diameter of work is 45 mm. depth of cut is 0.03 mm, stock is 0.6 mm for 220 mm long work piece with face width of wheel is 70 mm. **[6]**

    (b) Explain balancing of grinding wheel with neat sketch. **[6]**

2. **(a)** With the help of a neat sketch explain single point cutting tool geometry. **[6]**

    (b) In orthogonal cutting of a 60 mm diameter MS bar on lathe, the following data was obtained:

    Rate angle = 10°, Cutting Speed = 100 m/min, Cutting force = 200N, Feed Force = 70N, Chip thickness = 0.3 mm, Feed = 0.2 mm/rev.

    Calculate:

    (i) Shear angle, (ii) Coefficient of friction, (iii) Chip flow Velocity, (iv) Friction Angle **[6]**

3. **(a)** Define Tool life. Explain different factors affecting the Tool life. **[6]**

    (b) What are the important properties required for cutting tool materials? **[6]**

4. **(a)** Discuss about particle size, shape and size distribution and its effect on the properties of the final sintered compact. **[6]**

    (b) Discuss in general the process of powder metallurgy with respect to the following points **[6]**

        (a) Powder production,   (b) Compaction   (c) Sintering.

5. **(a)** With a neat sketch explain injection moulding. **[6]**

    (b) Write a note on forming and shaping of glass. **[6]**

6. **(a)** Write short note on Transfer Moulding and Blow Moulding. **[6]**

    (b) Write short note on : **[6]**

        (i) Transfer moulding

        (ii) Reaction – Injection moulding

        (iii) Rotational moulding

(P.1)

PAPER II

Time : 3 Hours　　　　　　　　　　　　　　　　　　　　　　　**Max. Marks : 60**

Instructions to the candidates :

(1) Each Question carries 12 Marks.

(2) Attempt any five questions to the following.

(3) Illustrate your answers with neat sketches, diagram etc., wherever necessary.

(4) If some part or parameter is noticed to be missing, you may appropriately assume it and should mention it clearly.

1.	**(a)**	Differentiate between Honing and Lapping process.	**[6]**
	(b)	Explain the meaning of each letter mentioned on the following grinding wheel.	**[6]**
		"W-C-10-E-5-V-17"	
2.	**(a)**	List the various types of chips produced during metal cutting.	**[6]**
	(b)	Explain theory of Lee and Shaffer.	**[6]**
3.	**(a)**	What is Machinability? Explain different factors affecting Machinability.	**[6]**
	(b)	State the different types of tool materials and explain how the appropriate tool material is selected.	**[6]**
4.	**(a)**	What are the different techniques used for compaction ? Explain Isostatic compaction in detail.	**[6]**
	(b)	Explain importance of sintering with its different stages to get required strength to green compact.	**[6]**
5.	**(a)**	Explain "Pressing" process of ceramics.	**[6]**
	(b)	Give the procedure of slip casting process.	**[6]**
6.	**(a)**	Explain in brief thermoforming process.	**[6]**
	(b)	Explain in brief Reaction Injection Moulding.	**[6]**

◈ ◈ ◈